T0260646

ADAPTIVE
ONCOGENESIS

JAMES DeGREGORI

ADAPTIVE
ONCOGENESIS
A New Understanding of How
Cancer Evolves inside Us

███ Harvard University Press
███ Cambridge, Massachusetts, and London, England
2018

Library of Congress Cataloging-in-Publication Data

Names: DeGregori, James, author.
Title: Adaptive oncogenesis : a new understanding of how cancer evolves
    inside us / James DeGregori.
Description: Cambridge, Massachusetts : Harvard University Press, 2018. |
    Includes bibliographical references and index.
Identifiers: LCCN 2017036331 | ISBN 9780674545397 (cloth)
Subjects: LCSH: Carcinogenesis. | Cells—Evolution. | Cancer—Etiology. |
    Cancer—Genetic aspects.
Classification: LCC RC268.5 .D437 2018 | DDC 616.89/1425—dc23
    LC record available at https://lccn.loc.gov/2017036331

To my sons and my parents

Contents

Illustrations by Michael DeGregori

Foreword

It has long been recognized that most adult human tumors—possibly all of them—arise through a complex, multistep process that usually requires decades to reach completion. This is most apparent at the level of histopathology, the science of examining tissue sections under the microscope. Histopathology reveals that many tissues throughout the body harbor benign, noninvasive growths and what appear to be more advanced growths, the latter being potentially aggressive and eventually life-threatening. Moreover, in some tissues, notably the colon, surgical removal of more benign growths greatly decreases the likelihood that fully formed, malignant carcinomas will arise in an individual. This proves directly that the cells in more malignant growths derive from those residing in more benign precursors.

These phenomena were placed in a genetic context in 1989, when it was revealed that the progressively more advanced growths contain more mutant genes in their genomes than their more benign counterparts do. Importantly, these mutations affected cellular growth-controlling genes, involving both oncogenes and tumor suppressor genes. Taken together, this suggested that a type of Darwinian evolution occurring in the microcosm of human tissues operated to govern the development of frankly malignant tumors. Thus, as premalignant cells acquire more and more mutant proliferation-promoting genes, their descendants acquire a type of selective advantage that allows such cells to outcompete neighboring preneoplastic cells. Such expanding clonal populations might, theoretically, expand until their size—perhaps millions of cells—makes it plausible that one of their number might acquire yet another randomly occurring mutation that conferred additional growth advantage, triggering a new round of clonal expansion. When depicted in this way, multistep tumor formation occurs as a consequence of a succession of clonal expansions that may, in many adult tumors, involve half a dozen or more of these steps.

These phenomena of selective clonal expansions directly parallel the evolution of species as first laid out by Darwin in the nineteenth century and then reinforced by the twentieth-century documentation of progressive genetic change, which came to be portrayed as the engine that powers multistep tumor progression. Unresolved by this model is the reason why human tumor formation is so complex. Here, a simple and likely correct notion comes to mind: that each of the steps of multistep tumor formation, achieved by one or another mutation in the genomes of preneoplastic cells, reflects the breaching of yet another barrier that the body has erected to prevent the outgrowth of malignancies. Such barriers may be erected by intracellular signaling circuits, by agents that preserve genomic integrity, and even by the actions of an immune system that strives in normal tissues to eliminate incipient cancers. According to this thinking, the more impediments to tumor formation, the less likely that a life-threatening cancer will appear in one's lifetime. Thus, if human tumor formation were hindered by only a small number of these barriers, we would all be covered with tumors by the age of two.

The extraordinary effectiveness of these multiple, concentric lines of defense in holding down cancer incidence is revealed by a simple fact: the cells in a human body undergo approximately 10^{16} divisions in a normal life span, with each cycle of cell growth and division representing an opportunity for a genetic disaster to occur—specifically, the types of disaster that generate cancer-promoting mutant genetic alleles. Indeed, only about 20 percent of people who die in our society are struck down by cancer, indicating that only about one in almost 10^{17} cell divisions serves to successfully spawn a lethal cancer.

This model of Darwinian evolution, as applied to cancer pathogenesis, is still incompletely proven although widely embraced. To be sure, there are certain details that distinguish the Darwinian evolution of a species in its native ecosystem from the evolution of preneoplastic cells occurring within tissues. This caveat aside, the Darwinian model seems robust but still underexplored in terms of its conceptual ramifications.

The present volume provides just such an exploration—a long-needed critical examination of how the work of Darwin in the mid-nineteenth century illuminates our twenty-first-century understanding of tumor development. Darwinian evolution is driven in no small part by changes in the surroundings of species that require novel adaptations, and indeed similar environmental changes occur within the interstices of the human body, in terms of alterations in tissue physiology—notably effects of nutrient availability, inflammation, tissue senescence, and immune function—that render individual cells more or less fit to proliferate and outcompete their neighbors. The arguments are carefully laid out, indicating why the present work will attract great interest among those who are interested in learning how cancers develop within the human body. This book represents a fascinating exposition of the parallels between these two types of evolution.

Robert A. Weinberg
Whitehead Institute for Biomedical Research
Massachusetts Institute of Technology
Cambridge, Massachusetts

ADAPTIVE
ONCOGENESIS

Introduction

WHY DO WE GET DISEASES like cancer, and why do we get them mostly when we are old? Why do children get cancer? Why do we age? These may seem like existential questions, but there are answers, even if we do not know all of them. If you go to your doctor, you will get proximate explanations, which consider the causes of processes and diseases like aging and cancer within us. You may be told that aging is the result of a lifelong accumulation of damage to your tissues, both from errors that occur as your body maintains itself and from the damaging exposures that we experience during life. Analogously, you will be told that cancer is the result of a lifetime's accumulation of genetic alterations (mutations) and that the accumulation by chance of a certain set of cancer-promoting mutations explains the increased risk of cancer late in life. Similarly, exposures to carcinogens, such as through smoking, are said to cause cancer by inducing mutations that change normal cells into cancer cells. But the more informative and more useful answers lie in an evolutionary understanding of life history and disease. As the great evolutionary biologist Ernst Mayr said, "No biological problem is solved until both the proximate and the evolutionary causation has been elucidated" (1982, 73).

Cancer is the second leading cause of death in industrialized countries: about 40 percent of people in those countries will develop cancer, and close to half of these individuals will die from their disease. Cancer is largely a disease of old age, with over 90 percent of cancers occurring in individuals over fifty. Other causes of cancer include cigarette smoking, alcohol consumption, sun exposure, pollution exposure, infections, and obesity. The medical and research communities have primarily ascribed associations between aging or exposures and cancer to enhanced mutation accumulation, despite strong evidence challenging such simple relationships.

This book will describe how the risk for cancer, like risks for other diseases, is inexorably tied to the life strategies that we and other animals evolved. These evolutionary strategies involve different investments in the maintenance of our tissues, resulting in very different potential life spans across the animal kingdom. Old age is not just the result of the passage of time; it is also the period when we experience physiological decline. The book will discuss the varied mechanisms that humans and other animals have evolved to avoid cancer, innovations required for the evolution of bigger bodies and longer lives. In particular, it will focus on how the maintenance of the body delays cancer until old age. Accordingly, this book will introduce a new theory: adaptive oncogenesis. This theory seeks to provide an explanation for why we and other animals are able to largely avoid cancer through periods of likely reproductive success. The theory postulates that animals have evolved stem cells that are well adapted to specialized niches within tissues (referred to as the microenvironment). Stem cells maintain a tissue for life, and thus they are the ultimate reservoirs of a tissue's genetic information for a lifetime. The maintenance of tissues during youth creates local environments that are unfavorable to change in the resident stem cells, including oncogenic changes that can contribute to cancer formation. In much the same way as stable environments on Earth limit selection for genetic change in organismal populations, healthy tissues suppress evolutionary change in constituent cells.

Unfortunately, humans and other animals do develop cancer. A primary reason is an unavoidable one—we get old. While the dominant explanation for the aging link is that substantial time is required for cells to accumulate sufficient numbers of mutations to cause cancer, this book will explore alternative explanations that are more in line with evolutionary theory. It will first consider how the force of natural selection to prevent cancer, other disease, or organ dysfunction diminishes in old age, as the odds of successful reproduction decline. Adaptive oncogenesis proposes that tissue decline during aging leads to selection for new cellular characteristics adaptive to this altered tissue landscape. Furthermore, the environmental exposures that our bodies experience can have substantial impacts on cancer risk, as clearly shown in the case of cigarette smoking. Drawing parallels with how changes in the environments of Earth have led to bursts of speciation, we will discuss how exposures like smoking perturb microenvironments for stem cells, leading to selection for oncogenic mutations that are adaptive in this damaged landscape.

This book will reevaluate clinical data within the adaptive oncogenesis framework, providing explanations for why cancer occurs in young children and why the detection of oncogenic mutations far outpaces incidence rates of cancers. It will discuss evolutionary theory–based strategies to modulate tissue microenvironments as a way to control the fate of precancer and cancer cells, and thus to prevent and treat cancers. The reductions in cancer mortality after almost half a century of the War on Cancer have been real but disappointingly inadequate. Greater focus on tissue landscapes and how they affect the evolutionary dynamics of both noncancer and cancer cells may be necessary to limit the pernicious impacts of this disease. All of these discussions will be framed in the light of evolution, as urged by the great evolutionary biologist and geneticist Theodosius Dobzhansky (1973). For better or worse, we and all other animals are products of the forces that shaped our evolution. Thus, we will start with an understanding of life histories among different animals, and how and why they evolved.

CHAPTER 1

The Evolution of Life Spans and Disease Avoidance

LIFE MAY SEEM cruel and capricious. Babies can be born with developmental defects, some of which can lead to early death and others of which can engender lifelong disabilities. Newborns and very young children can develop cancers, the most common of which is leukemia. Our lives are also limited. As we age, we experience the loss of our vigor and strength, and at times even our mental functions, as our bodies undergo their inevitable decline. In old age, we become more susceptible to diseases, including infections, heart disease, Alzheimer's disease, and cancer. These painful facets of life seem hard to explain. Life appears even more cruel for animals in the wild. The hazards they face are many, diseases are common, food is scarce, and the weather can be deadly. One animal might be another's next meal, and if that predator does not find its next meal, it will likely perish. So no matter where an animal is on the food chain, life is very risky. We can make sense of these facts of life by understanding how evolution has sculpted life strategies, with the overarching goal of maximizing reproductive success. When and why humans and other animals develop diseases such as cancer has been influenced by the evolutionary history of each species and the environments that selected for particular life strategies.

Of Mice and Malthus

First published in 1798, Thomas Malthus's *An Essay on the Principle of Population* described how human population growth is restrained by resource limitation, with exponential growth restricted by famine and disease. In the 1800s, Charles Darwin and Alfred Russel Wallace both recognized the implications for understanding species evolution: many if not most animals of a given species will die before reproducing or otherwise not reproduce successfully. If not, the world would quickly fill up with this species.

As an example, we do not need to study mice in the wild to know that most of them die early. We just need to know that the house mouse (*Mus musculus*), for example, can reproduce within about six weeks of birth, that its average litter size is around ten, and that gestation lasts less than three weeks. If all mice survived and reproduced, a pair of mice and their progeny would generate about 360 million mice in one year, assuming unlimited resources and no predators. While this may appear unbelievable, one need only consider that every three weeks ten pups would be born, which would contribute five more male-female pairs, which themselves would be able to mate within only six weeks of life. Since we are not up to our noses in mice, we can safely assume that the vast majority of mice never even successfully breed and likely die at a very young age. For those that do produce pups, most of these offspring will die before having pups of their own.

Why does any of this matter for understanding aging, disease, and cancer? We are products of our evolutionary history. Evolution works to maximize reproductive success: everything else in biology is just details. When we consider aging and cancer and other diseases in the context of the evolutionary forces that shaped us, we can derive certain first principles upon which all other details about aging and disease can be based. The first of these principles is that aging—the breakdown of our tissues and the decline in overall physiology—has been determined by selective pressures for life histories adaptive to environmental conditions. A life history for a species describes parameters such as body size and the timing and character of maturation, reproduction, and aging that evolved to maximize adaptation to a particular environment. Even within the same environment, two different species occupying different niches can adopt very different strategies for reproductive success. In the same grassy plain, the field mouse and the bison will live very different lives, and should they be lucky enough to live a long life, they will age at very different rates. First, we need to consider how these different life strategies evolve.

Natural Selection for Different Life Strategies

Darwin and Wallace described the process of natural selection, whereby selection from a population of individuals with diverse traits acts to maximize fitness. Fitness is a measure of reproductive success—the tendency of individuals to decrease or increase in frequency within a particular environment dependent on their heritable characteristics. Fitness is determined by the genetic makeup of the organism, or the genotype. The effect on fitness for any genotype will be highly dependent on environmental context. Different genotypes will be manifested as different traits and characters of the organism, known as phenotypes. Within an environment that has been stable for a long time, a population will have evolved through natural selection to maximize fitness within this environment. The status quo will be largely maintained, as genetic changes that lead to phenotypic changes will typically reduce fitness. It is hard to improve upon an already well-adapted state. Should the environment change, there will be selection from within the diverse genotypes present in the population for individuals that are better adapted to the new environment. An example is antibiotic resistance in bacteria. The genotype conferring antibiotic resistance is not advantageous and in fact comes with a cost in the absence of antibiotic treatment (Melnyk, Wong, and Kassen 2015). However, when we are treated with antibiotics, the bacteria bearing this same genotype suddenly become much more fit than their antibiotic-sensitive brethren, and they will quickly dominate the population—an example of positive selection. Both selection for and against the antibiotic-resistant genotype are part of natural selection. As an example from our own species, the plague (*Yersinia pestis*, often called the Black Death) killed millions of Europeans in the fourteenth century, leading to selection for new variants of immune genes that provided protection from *Y. pestis* (Laayouni et al. 2014). By better surviving the Black Death, people carrying these gene variants had better odds of reproducing, and thus the frequency of these genes increased in Europeans following the strong positive selection created by the plague.

Natural selection has acted over millions of years to generate a diversity of animals, with millions of different strategies for reproductive success. Sometimes this strategy is to live fast, produce lots of offspring, and die young. This fast strategy is typically selected for in a highly hazardous environment. The field mouse would be an example of an animal using the fast strategy. Field mice constantly face external hazards, from predators to pathogens and from scarce food to low temperatures. Thus, natural selection has not invested in the long-term maintenance of mouse bodies (aging starts by about one year of life), and mice breed as early and often as they can, with large brood sizes. Nevertheless, as described above, most mice will be unlucky, dying without leaving offspring.

The slow strategy is to mature later, invest more in a smaller number of offspring, and live longer. Natural selection favored this strategy in humans, elephants, tortoises, and blue whales, to pick a few disparate examples. In each case, characteristics of these species—such as intelligence for humans, large size for elephants and whales, and a hard shell for tortoises—has reduced external hazards, and thus investment in longer lives can provide a return in the form of additional offspring for chronologically longer times. Natural selection unconsciously acts like a financial investor: investment is made to the point that a return on the investment is likely. For the mouse, investing in body maintenance past one year would be unlikely to pay off in terms of offspring (which is the only thing that matters in the end). But for a blue whale or an elephant, which experiences minimal predation after reaching maturity, investing in longer lives can pay off with additional offspring.

Evolutionary biologists have appreciated the impact of life strategies on aging for more than half a century: the decline in our tissues (collectively called the soma) with age has been postulated to reflect the lack of selection for tissue (somatic) maintenance beyond the age when the animal contributes to population fitness by passing its genes on to future generations (Kirkwood 2005). Peter Medawar described this idea in his famous 1951 lecture at University College London: "The force of natural selection weakens with increasing age. . . . This dispensation may have a real bearing on the origin of innate deterioration with increasing age" (1952, 18).

Medawar proposed that if an inherited (germline) mutation occurs that reduces survival probability, the cost of the mutation depends on the age when this cost is experienced. If a mutation decreases survival before sexual maturity, it will have a large cost and will be strongly selected against. However, if a mutation decreases survival only in the very old (for example, promoting neurodegeneration), at ages well past when reproductive success is likely, then the impact on fitness will be minimal to nonexistent. These concepts were mathematically formalized by William Hamilton (1966), who quantified why mutations earlier in life have a much greater impact than those later in life. In fact, for humans it is estimated that the fitness effect of a phenotype-altering mutation at the ages of forty and fifty years are one-fifth and one-twentieth, respectively, of the effect of a mutation at twenty—whether this effect is positive or negative (Tuljapurkar, Puleston, and Gurven 2007). Furthermore, if an inherited mutation has a benefit early in life but comes with a cost later in life, the balance of benefits and costs will be weighed by natural selection. George C. Williams (1957) proposed that this antagonistic pleiotropy of mutational effects contributes to aging. Phenotypes that benefit reproductive success early but promote aging later in life have likely been positively selected throughout evolutionary history. Given the many hazards of life, the physiological decline associated with old age is not thought to be a major factor limiting reproductive success for wild animals. As described by Medawar, "animals do not in fact live long enough in the wild to disclose the senile changes that can be made apparent by their domestication. . . . It is of vital importance to remember that senility is in a real and important sense an artifact of domestication; that is, something revealed and made manifest only by the most unnatural experiment of prolonging an animal's life by sheltering it from the hazards of its ordinary existence" (1952, 13). Notably, the aging-associated physiological decline known as senescence is actually observed in wild animal populations, but typically in just a small fraction of these animals, and the odds of reproductive success are slim for these individuals relative to the younger members of their populations (Kowald and Kirkwood 2015).

For similar reasons, cancer appears to be rare "in the wild," in line with what Medawar noted for senescence (Hochberg and Noble 2017). Cancer does not appear to be a major cause of mortality for wild animals in a truly natural environment (Figure 1.1), although malignancies could still contribute toward reducing survival for a fraction of animals (Madsen et al. 2017; Roche et al. 2017). Nonetheless, cancer risk increases dramatically at ages where most wild animals are unlikely to survive. A mouse that lived long enough to have a high risk of cancer would be a lucky mouse indeed. For all animals, there has been strong selection to avoid cancer during periods of likely reproductive success, and thus death from external hazards such as predators, disease, cold, and starvation is far more likely than death from cancer or other diseases of old age.

For humans, natural selection has been relatively blind to the elderly, which during most of our evolutionary history represented a much smaller fraction of our population than today (Kirkwood 2005). The chances of a human older than about fifty contributing to the genetic makeup of future generations were low, as an earlier demise due to disease, starvation, predators, or other causes was more likely. Clearly, modern humans in developed nations are living under conditions very different from those of our ancient ancestors. As described by Randolph Nesse and George C. Williams in *Evolution and Healing: The New Science of Darwinian Medicine* (1996), an average hunter-gatherer woman might have four children, two of whom would be likely to die before reproducing—which explains the fact that the human population remained relatively stable in size for the 70,000 years preceding the adoption of agriculture. In the Stone Age, "death always balanced reproduction" (ibid., 136). In fact, fossil evidence indicates that 50,000 years ago only roughly 10 percent of our African ancestors lived to the age of forty, and less than 5 percent reached fifty (Kvell et al. 2011). Additional evidence indicates that survival into older adulthood was highly improbable up to about 50,000 years ago (Caspari and Lee 2004).

Still, the longevity of our ancestors continues to be debated. In contemporary hunter-gatherer populations, an individual has a reasonable chance of living past forty-five if he or she survives to adulthood, with

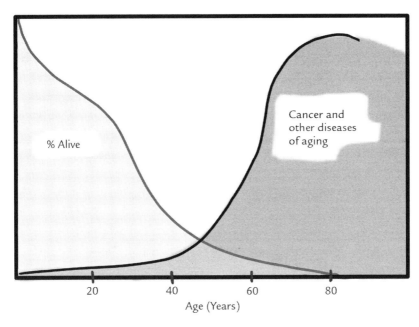

Figure 1.1. Animals in the wild rarely suffer cancer and other diseases of aging. Given external hazards and resource limitations, most animals in the wild do not survive to ages where cancer and other diseases of old age (like heart disease or neurological disorders) become common. Natural selection has acted to postpone senescence and these diseases to periods when reproduction is less likely. The curves shown are for humans. The "% Alive" curve shows estimated survival for Europeans about 15,000 years ago (Kvell, K., et al., Chapter 1. "Gerontology," in *Molecular and Clinical Basics of Gerontology*, 2011), and the "Cancer and other diseases of aging" curve approximates cancer incidence for Americans in 2010 (graphed from data tabulated at www.seer.cancer.gov). © Michael DeGregori

about half of those surviving to twenty making it to sixty (Kaplan et al. 2000; Gurven and Kaplan 2007). Notably, these studies suffer from the caveat that survival in these populations has been affected by contact with the developed world. Regardless, we can appreciate that reproductive success requires survival of a parent well beyond childbirth—the death of either parent before the child reaches maturity would greatly reduce offspring survival probabilities. Moreover, researchers have surmised that selection for human longevity may be somewhat extended, given the so-called grandmother effect: grandparents could facilitate the survival of their grandchildren and thus their genetic lineage by helping with their care (Brown and Aktipis 2015). These contributions to future generations may explain why humans typically maintain good health well into their fifties (see Gurven and Kaplan 2007), as evidenced by low rates of death through these decades in modern wealthier nations (see Chapter 7). Mortality rates rapidly accelerate after age sixty in both modern hunter-gatherer societies and in developed nations. Nonetheless, selection to avoid aging, cancer, and other diseases in the elderly would still be weaker than in younger individuals, given low odds of surviving to older ages for most of human evolution, and their contributions as grandparents to offspring survival has been likely less than for a parent (Tuljapurkar, Puleston, and Gurven 2007). Moreover, we can speculate that grandparenthood may have started in the forties (assuming parenthood began by age twenty in ancient times), and thus selection for fitness in what we now consider old age would have still been weak. Obviously, each individual human has a greater chance of living to age n years than to age $n + 1$, so odds for survival declined significantly with each year of life in ancient times.

The development of agriculture and animal domestication led to a gradual rise in the human population over the past 10,000 years, facilitating the development of large civilizations. The greatest population explosion for humans occurred in the last 200 years, with the population growing from about 1 billion in 1800 to 7.4 billion today (Bongaarts 2009). In parallel, average life expectancy more than doubled in developed countries, from less than thirty-five years in 1820 to more than eighty in the twenty-first century. Changes in life expectancy were largely driven by improvements in hygiene, medicine, nutrition, and public health, with much of the impact on average survival resulting from reduced mortality for the very young. For the vast majority of human evolution, living to our golden years was the exception, not the rule. The situation is rapidly changing. In 2010, 524 million people, or about 8 percent of the world's population, were sixty-five or older. In line with reductions in mortality combined with declining birthrates, this number is expected to triple to 1.5 billion by 2050, with 16 percent of all humans over sixty-five. With extended life spans, diseases of old age—from heart disease to Alzheimer's disease and cancer—are affecting far more people.

How will these changes in human demographics affect the selective forces acting on our species? Will natural selection favor delayed aging now that humans live longer, and in some cases reproduce later in life? The answer is, probably not any time soon. First, having one's first child after thirty may be common in more educated communities in developed nations, but it is less prevalent in much of the world. Second, it would take many generations of humans reproducing at older ages to alter the evolved life strategy, and that would happen only if such a strategic shift improved reproductive success—which includes offspring survival. In modern society, even if a parent dies after age fifty, offspring already born will almost certainly be cared for and survive.

Aging: The Waning of the Force of Natural Selection

The frailty and physiological decline of old age, for humans and other animals, is a reflection of the substantially reduced probabilities for older individuals to contribute to the successful propagation of genes to subsequent generations. Cancer risk similarly increases as the odds of successful reproduction decline. The sad truth is that there has been minimal selection against cancer and other diseases of the elderly, as preventing these diseases and conditions would have required additional investment during youth. It is better to invest in ensuring robustness and reproductive success in youth. Evolution essentially plays with probabilities: selection will favor a strategy of investment in tissue maintenance that maximizes the return on this investment, and if the odds of successful reproduction in a population are small past a certain age, selection will not invest in avoidance of diseases, including cancer, past this age.

Tissue decline and cancer are not the products of evolved strategies to eliminate the elderly to make room for the young. We and other animals have not evolved a program to eliminate older individuals, so that resources can be better used by our offspring (Kirkwood 2005). First, if we had, evolution would quickly select for cheaters who did not have the aging program, as they would continue to produce offspring at later ages than the more compliant members of the population. Second, natural selection simply becomes very weak at ages when most individuals would be dead by other causes: there cannot be much selection for a program to eliminate individuals at ages when few individuals would be around. Aging and cancer in old age are not the product of an evolved program, but indicate the waning of evolved programs to prevent somatic decline and disease.

We've talked about investments in youth to maximize fitness or reproductive success, but what are these investments? More importantly, how are they modified by evolution to forge different strategies for life? The short answer is that we do not know. However, we do have clues.

Presumably, if there is an investment, it must have a cost—for if it had no cost, evolution would favor the extension of longevity even if it improved the odds of reproductive success only for the rare individual. The most logical cost is energy, and for animals energy must be obtained in the form of food. Nutrients in food such as sugars, fats, and proteins provide the fuel for all of our bodily functions and the building blocks for us to develop, grow, and be maintained. With the notable exception of recent times for humans, food is typically limited—whether for a predator that has to chase down its next meal, or an herbivore trying to survive the winter. Food can be difficult and risky to obtain, and it can take a lot of time to find, consume, and digest. Thus, the energy and building blocks derived from food need to be invested judiciously.

Considering this last word (judiciously), we need to keep in mind that natural selection does not have foresight, makes no calculations, and does not think. Natural selection simply works with outcomes, representing a blind process of differential survival and reproductive success among a variety of natural phenotypes. If a mutation occurs that is advantageous, it is more likely to persist and expand in the population. Thus, if a mutation occurred that led to greater energetic investment during youth that would delay physiological aging, then this mutation would be "judged" by natural selection to determine whether the added investment paid off in terms of more surviving offspring. In an environment that has been stable for a long time, this added investment is likely to be selected against, since the proper strategy has already been optimized for this environment. Increasing somatic maintenance in youth to delay aging would come with a cost in reproductive success during youth, when that success is most likely. If energy is indeed the currency, we can see how investing this energy in youthful vigor would be more profitable than in maintaining the bodily robustness of a relatively few older individuals.

Let us now consider what would happen if the environment changed. What if the numbers of predators dropped substantially, or other external hazards were reduced? The same mutation delaying aging could be adaptive (increasing fitness), as reduced external hazards would increase the odds that individuals would be around to mate at older ages. Robustness (and less frailty) at these older ages, which would come with the cost of added investment in youth, would now pay off in terms of increased overall reproductive success. Thus, we can see how natural selection tunes somatic maintenance strategies to optimize reproductive success. As George Williams noted in his famous paper on the evolution of senescence: "There are therefore, two opposing selective forces with respect to the evolution of senescence. One is an indirect selective force that acts to increase the rate of senescence by favoring vigor in youth at the price of vigor later on. The other is the direct selection that acts to reduce or postpone the 'price' and thereby decrease the rate of senescence. The rate of senescence shown by any species would depend on the balance between these opposing forces" (1957, 402).

An example of how alterations in the environment can quickly modify longevity comes from Steven Austad's studies of the Virginia opossum, *Didelphis virginiana*, which lives in the southeast of the United States. Austad (1993) studied a population of these opossums on Sapelo Island off the coast of Georgia. This island has been separated from the mainland for only about five thousand years. This separation led to a great reduction in the numbers of predators for opossums. It had previously been shown for mainland populations that over half of recorded opossum deaths can be attributed to predation, mostly by other mammals such as bobcats and canines. The evolutionary theory described above would predict that the island population, with reduced external hazards, would age more slowly given greater odds of later reproduction. Indeed, by measuring multiple physiological parameters, including reproductive abilities and tail characteristics, Austad demonstrated that, relative to the mainland population, the island opossums exhibit a rate of aging that is reduced by more than 1.5 times. It is important to point out that opossums demonstrating signs of physiological decline (senescence) are apparent in a fraction of both the island and the mainland populations, but the hallmarks of senescence appear at older chronological ages for the island opossums. Reductions in predation on the island appear to have promoted selection for delayed aging, as opossums on the island have a greater chance of successfully producing offspring in both their second and third years of life, while similar success is less likely on the mainland (Figure 1.2). Not only is physiological aging delayed, but the island opossums exhibit another manifestation of a slower life strategy: smaller brood size. This smaller size could reflect a cost of delayed senescence (energy conservation for later in life) or a better reproductive strategy via greater investment per offspring. Finally, Austad examined other factors, such as weather, food availability, and infections, that might account for these differences in life history. None of the factors showed differences that would indicate nonevolutionary causes of delayed senescence in island opossums. What is amazing is that within an evolutionary blink of the eye—just five thousand years—such a dramatic change in somatic maintenance strategies can evolve.

Austad's study illustrates the power of environmental change (in this case, predation) to alter selective pressures, leading to the evolution of a new strategy. Moreover, it indicates that the programs controlling the rate of aging are pliable. If evolution can tune the rate of aging, can we do so artificially? We will return to this idea in Chapter 12.

On a broader scale, vastly different life spans have evolved within a single order, *Rodentia*. The naked mole rat, *Heterocephalus glaber*, lives up to thirty years, while the house mouse, *Mus musculus*, can survive only to around three years in captivity (and would probably not live even one year in the wild) (Buffenstein and Jarvis 2002). Naked mole rats live in sealed underground burrows and are thus protected from many of the usual hazards of living aboveground. The house mouse may experience the joys of fresh air, but each day comes with great risks, from starvation to predation. Thus, the timing of the onset of aging is primarily determined by environmental pressures that select for life history strategies maximizing reproductive success.

We need to consider risk in terms of probabilities. Just as in life it is impossible to eliminate all risk, natural selection works to maximize an individual's odds of reproduction. Natural selection will only invest to the point that doing so benefits reproductive success, and this calculus entails some risks of premature mortality—whether through predation,

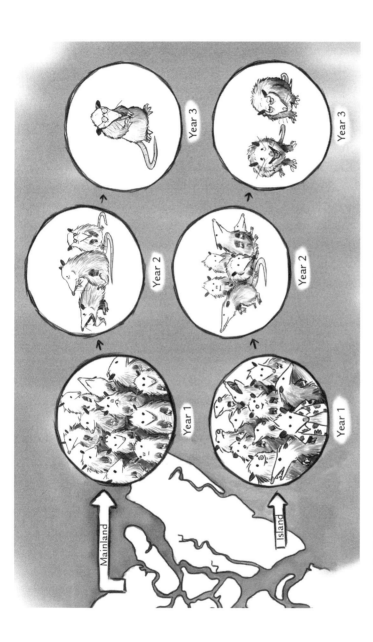

Figure 1.2. Delayed senescence evolved for island opossums. Mainland opossums, shown in the top circles, experience greater predation, and thus natural selection did not act to prevent physiological aging at ages when reproductive success was unlikely (in year two of life and beyond). Island opossums, shown in the bottom circles, experienced less predation over the last five thousand years or so, selecting for prolonged tissue maintenance programs given reasonable odds of reproductive success in year two. Signs of aging are thus delayed to year three for the island opossums. © Michael DeGregori

environmental causes, or disease. If the cost of further reductions in risk outweighs any potential increases in the chances of successful reproduction (the return on the investment), then natural selection will not favor those reductions. Cost-benefit determinations will depend on the current environment and context. Potential reductions in risk are not limited to old age, as strategies to limit disease and other hazards in youth also come with trade-offs, not to mention the constraints of evolutionary change.

We can conceptualize cancer within this same evolutionary framework. Selection against cancer has been strongest in the very young, as developing cancer would likely ensure that the individual would not reproduce. Not only does cancer frequently lead to the death of an individual, but the burdens of the disease could lead to even earlier demise in the wild, such as through reduced ability to avoid predation (Roche et al. 2017). With each year after an individual reaches sexual maturity, the strength of this selection against cancer diminishes, as the odds of reproduction are reduced.

The callous reality of evolution is that natural selection acts to promote an individual's health and longevity only to the extent that doing so benefits reproductive success. As Williams concisely notes, "natural selection may be said to be biased in favor of youth over old age whenever a conflict of interests arises" (1957, 401). We have to remember that selection across generations does not directly act on the soma (the body). The soma is a vehicle for gametes (sperm and eggs) that evolved to maximize the odds of successful reproduction. Nonetheless, the evolution of the entire corporeal plan—the cells and tissues that make up each body, and the strategies employed to form and maintain the body—and its adaptation to the exterior environment are driven by selection at the germline level. In particular, the organization and behaviors of somatic stem cells, charged with maintaining tissues throughout an animal's life span, and their dependency on specialized niches in tissues have been honed over millions of years of animal evolution. These niches include other cells, the extracellular matrix, soluble growth factors, and blood vessels. Since transgenerational selection acts on the genetic information passed along in gametes (the germline), we will refer to it as germline selection. Germline selection promotes the evolution of a soma that best achieves the transfer of genes, typically involving high bodily fitness through periods when reproduction is most likely, with a waning of this fitness as the odds of reproduction decline. In this sense, the soma is disposable (Kirkwood 2005).

In the production and maintenance of animal bodies composed of up to trillions of cells, mutations occur and selection can act on these somatic cells to change the proportion of cells with particular genotypes, thereby affecting the cells' somatic fitness (Rozhok and DeGregori 2015). We refer to this process as somatic evolution. Somatic fitness is a measure of the success of a cell of a particular genotype in competition with other cells. Some forms of somatic evolution, typically involving multiple rounds of selection for cells with mutations that improve somatic cell fitness, can lead to cancer. The evolution of different life strategies across the animal kingdom will not only lead to different rates of aging, but also to correspondingly varied patterns of cancer incidence. In the following chapters, we will learn how tissue maintenance in youth limits cancer development by favoring normal cells, and how deteriorating tissue microenvironments resulting from physiological aging lead to selection for oncogenic mutations that allow somatic adaptation to the new conditions. Evolved patterns, such as delaying cancer till old age, can be disrupted by lifestyle choices and the exposures inherent in modern living, through alterations in tissue landscapes. Thus, both processes we can control (such as whether to smoke) and those we cannot (our evolutionary histories) can profoundly affect the evolutionary trajectories of somatic cells and cancer risk.

CHAPTER 2

Understanding Evolution at Organismal and Somatic Levels

CHARLES DARWIN AND ALFRED RUSSELL WALLACE'S theory of natural selection was not readily accepted by biologists after Darwin's publication of *The Origin of Species* (1876) in 1859 (preceded by the publication of papers by both scientists). Natural selection lacked a mechanism, whose identification needed to await the rebirth of genetics in the early 1900s. There was also a paucity of observational or experimental demonstrations of natural selection in action until the twentieth century (Mayr 1982). The mutation theory of Hugo de Vries and William Bateson dominated the beginning of that century, advocating that evolution proceeded by depending on mutations (typically those with large effects, known as saltations) rather than on selection. Mutation alone was thought to be sufficient to bring about evolutionary change, with any role of selection firmly dismissed. This view was particularly dominant among geneticists. Mutation-driven evolution was rejected by naturalists, who by neglecting the importance of genetic variation also rejected natural selection. A revolution in biology and evolution required new evidence and new minds, which led to the development of the evolutionary synthesis (also called the modern synthesis) between 1936 and 1947. The evolutionary synthesis revived the theory of natural selection and incorporated population genetics into our understanding of evolutionary change. Its proponents stressed gradual evolutionary change and argued that mutations by themselves were directionless, with the action of selection determining directional phenotypic change. Geographic isolation was proposed as a key mechanism for speciation. Ernst Mayr (1982) describes the frequent misconception, throughout the modern history of biology, that genes and mutations are the targets of selection. Instead, the whole organism is the target, and the genotype-determined phenotype of an organism will be subject to selection under the demands of its current environment, including competition with members of the same species. This book proposes that the same rules apply to somatic cells: the fitness of a genotype is realized only in the context of tissue environment. A new mutation is judged by natural selection in the context of the entire genotype of the cell, the state and number of competing cells, and the current microenvironment.

Selection, Mutation, and Drift

Evolution is sculpted by three fundamental forces: mutation, drift, and selection. The first two are essentially random and not inherently creative. The third, selection, is the creative force. Selection works on the diversity of morphological, biochemical, and functional types (phenotypes) that exist in a population, which were brought about by the processes of mutation and drift. Drift represents the random assortment of genetic types. It is important to point out that selection is not seeking some goal, nor is the inevitable outcome greater complexity, intelligence, size, or any other property that we might consider better. Selection works to maximize fitness within the current environment, and when the environment changes, selection will again work to promote adaptation of the organism to the new environment. Thus, unlike drift and mutation, selection can be deterministic (Lipinski et al. 2016). Selection can promote change in a particular direction.

A mutation is any change to the genetic material—DNA for all life, except some viruses. A point mutation is a change in a single base, altering one letter in the code made up of four bases (A, T, C, and G). Such changes can result from an error during DNA replication. High-fidelity DNA-replicating machines (polymerases), together with very effective

DNA repair machinery, greatly limit the generation of mutations. Since DNA replication has a low error rate, and DNA repair is very efficient, mutation rates are generally kept very low (around 10^{-8} to 10^{-10} mutations per base per cell division) (Lynch 2010). Since any particular amino acid is encoded by a set of three consecutive DNA bases, a single letter change will often change this code, and thus the mutated gene will encode for a different protein. Importantly, this alteration in the protein can affect function, and this impact can range from inconsequential (minimal to no change in function) to very large (greatly affecting function, for example by increasing or decreasing the corresponding protein activity). Mutations can also affect regulatory sequences that control gene expression and genes that encode functional RNAs. RNAs can participate in gene regulation, protein production, and other cellular roles. Mutations can also be much larger, including substantial deletions within chromosomes, duplications of parts of chromosomes, gains or losses of whole chromosomes, and rearrangements (Hoeijmakers 2001). Rearrangements include inversions within a chromosome or translocations in which a piece of one chromosome joins with a broken piece of another—which can in turn generate novel fusion proteins that are hybrids of two different genes.

Mutations typically result from an error in DNA replication or damage inherent to cellular living, but they can also result from external insults like cigarette smoking or exposure to ultraviolet radiation from the sun (Hoeijmakers 2001). Mutations are pretty rare, but they occur at a high enough frequency to fuel the evolution of species as well as that of cancers. I will often refer below to oncogenic mutations, which are mutations that have the potential to contribute to the generation of cancer phenotypes. I will discuss how this potential is context-dependent. Oncogenic mutations can include gain-of-function genetic changes that amplify or alter the activity of the mutated gene, as well as loss-of-function changes, such as those mediated by chromosomal deletions, that reduce or eliminate gene activity (Hanahan and Weinberg 2011). Genes subjected to gain-of-function oncogenic mutations are called oncogenes, and those subjected to loss-of-function oncogenic mutations called tumor suppressor genes.

Heritable variation in a population is necessary for evolution. But how do we get form and function from variability caused by random mutations? The answer of course is selection. Heritable variation is the substrate upon which selection acts. We can borrow concepts from the field of population genetics, founded by Ronald Fisher (1930), J. B. S. Haldane (1932), and Sewall Wright (1931). Fisher's early formulas demonstrated that selection is proportional to the variation present in the population, which itself is typically proportional to the size of the population. Within our bodies, we can consider that the populations of interest are the cells that have some capability of becoming cancers. These definitely include stem cells, as well as some additional cells that maintain the ability to divide (those that have not terminally differentiated). Fully functional cells of a tissue, such as muscle cells, neurons, red blood cells, and skin epithelial cells, are called differentiated.

Since stem cells are the precursors of all cells of a tissue and continue to repopulate the tissue with new cells during the life of the organism, these cells are also the ones thought to be most susceptible to being transformed into cancer. Nonetheless, cells with shorter life spans that still make other cells (referred to as progenitor cells) can also be targets for cancer formation (oncogenesis). There is also evidence that progenitors and even differentiated cells can be reprogrammed to become stem-like cells, particularly when under stress (Wahl and Spike 2017). Cell divisions, when mutations mostly occur, create risk for cancer. Thus, by maintaining their ability to divide, stem cells and in some cases progenitor cells can accumulate the mutations that are the fuel for somatic evolution. Importantly, as I stress throughout this book, simply accumulating mutations is not enough to lead to cancer. Whether potentially oncogenic mutations are selected for in the somatic cell clone, thus leading to a cancer, is highly dependent on context.

While we often envision selection as being positive, selecting for a particular phenotype, most selection is negative. Negative, or purifying, selection is the process that eliminates disadvantageous mutations from the population. Just as randomly changing a wire in a radio is likely to impair the function of the radio, most mutations in either a somatic cell or an organism are likely to be disruptive. Purifying selection works to eliminate mutations from the population of cells or organisms. Some mutations, such as those that disrupt essential cellular functions like DNA replication, will always be selected against. More commonly, whether a mutation is positively or negatively selected will depend on context.

For example, for somatic cells, a mutation that confers resistance to challenges imposed by low oxygen (hypoxia), but otherwise comes with some cost, will be negatively selected in a well-vascularized tissue with physiological oxygen levels. But in a tissue experiencing hypoxia, which can occur during tumor growth, this same mutation would be positively selected. To put this further into evolutionary parlance, the allele frequency of the gene with the mutation that confers hypoxia resistance would increase in the hypoxic tissue or tumor, as there will be more cells bearing that allele. Alleles represent different versions of the same gene, with often minor differences in the DNA sequence and the encoded protein. Both selection for and against mutated alleles are critical components of somatic evolution.

These rules for how selection affects mutation and allele frequency hold well for large populations. But what about small populations? And what happens to mutations that have little to no impact on phenotype? In these contexts, genetic drift becomes important. Drift is the role of chance in the fate of a genetic allele. According to the classic population models of Fisher (1930), Haldane (1932), and Wright (1931), the power of drift is inversely proportional to population size. In small populations, drift can significantly diminish the effects of selection. If we flip a coin ten times, we would not be too surprised if seven or more of the flips gave us heads. In fact, the odds of such an outcome are about one in six. However, if we flipped the same coin a thousand times, the odds of getting heads seven hundred or more times would be less than one in a million. The greater the number of trials, the smaller role chance plays.

The same principles apply to an animal or any other organismal population, as well as to somatic cells. For example, drift can be important when a population experiences a bottleneck, such as when a few migrating animals initiate a new population in a new location—like the first iguanas to land on the Galapagos Islands. For better or worse, the genetic alleles that came with these founding iguanas were fixed in this new population. This does not mean that selection plays no role in such contexts, as a mutation with a large effect on fitness would still be subject to selection. Fisher's formulas reveal a simple relationship: if a mutation has an effect much greater than the inverse of the population size, then its frequency will be largely influenced by selection. Therefore, the fate of a mutation that increases or decreases fitness by 1 percent in a population of ten thousand will be largely dependent on the selective value it confers. The converse is also true: if a mutation has a substantially smaller effect than the inverse of the population size, its frequency over time will be largely dependent on drift. Thus, the fate of a mutation that increases fitness by 0.1 percent in a population of one hundred will be mostly due to drift. Of course, for all genotypes that affect fitness, both drift and selection play some role, but their relative contributions depend on population size.

Following these rules, one can surmise that drift will also be important for genetic changes that have essentially no impact on fitness (known as neutral or nearly neutral mutations), regardless of population size. In fact, most of the evolution of the genomes of vertebrates is under drift, as most changes have no or only negligible influence on phenotype. Only 1–2 percent of our genome encodes for proteins, and even considering regulatory sequences and genes encoding for important RNAs, at least 90 percent of our genome appears to be under no or very weak selection and is thus largely if not entirely subject to drift (Lindblad-Toh 2011). The reason for this discussion of drift is that it plays an important role in the fate of both organisms and somatic cells with mutations, together with the important role played by selection. The same three forces—mutation, selection, and drift—that determine organismal evolution are also critically important for somatic evolution and thus cancer.

Navigating Fitness Landscapes

The adaptive value of a new mutation will depend not only on the environmental context, but also on the genetic context of the organism or cell in which the mutation occurs. For example, if a cell previously acquired phenotype-altering mutations, then the fitness of this cell and its progeny (cellular "clones") will be determined by the complete genetic picture, including both new and old mutations. For this reason, the order in which mutations occur is important for adaptation. Daniel Weinreich and colleagues (2006) showed that the path to the development of strong antibiotic resistance in bacteria, mediated by the β-lactamase gene with five point mutations, was unlikely to occur through all but 18 of the possible 120 mutational paths ($5 \times 4 \times 3 \times 2 \times 1 = 120$). For example, if mutation C increases antibiotic resistance (and thus fitness) only when it occurs in a bacterium that already has mutations A and B, mutation C would not be likely to persist or confer an advantage to a bacterium in the absence of both other mutations. The same relationships are observed in cancer evolution, where particular oncogenic mutations are found to co-occur with particular other mutations, to accumulate during tumor evolution in a particular order, and sometimes never to occur with certain other mutations (Yates and Campbell 2012). Essentially, if evolution must proceed uphill in terms of fitness, then a mutation on this path needs to confer a phenotype that improves fitness.

This brings us to fitness landscapes, also referred to as adaptive landscapes. Wright (1932) first envisioned evolution in terms of landscapes, which are topological maps of possible evolutionary trajectories for an organismal population. We will also use fitness landscapes in discussing somatic cell evolution. As can be seen in Figure 2.1, possible genotype-encoded phenotypes are arrayed on the X-Y plane, and the height represents the fitness of these phenotypes. A given graph will represent these relationships for a particular environment. Evolution will mostly proceed up a peak and should not progress downward—that is, selection will act to increase fitness. These rules for evolutionary trajectories on a fitness landscape as described above hold well for large populations, whether organismal or somatic, and for mutations that have consequential phenotypic effects.

Contrasting Germline and Somatic Evolution

Our bodies can be conceptually considered as vehicles for the germline, and natural selection has acted to maximize the capacity of these vehicles to contribute genetically to subsequent generations. From this perspective, all somatic cells are subservient to the body. The only evolutionary strategy that ultimately matters is the organismal one: maximizing reproductive success. Cancer cells will be less compliant than normal somatic cells, reducing the fitness of the individual if they occur early enough in life.

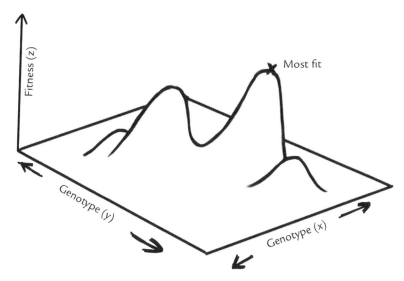

Figure 2.1. Hypothetical fitness landscape. Topographical depiction of the fitness of possible genotypes for a population in a particular environment is shown. Potential genotype-encoded phenotypes are arrayed on the X-Y plane, and their proximity indicates their genetic relatedness. The height (the protruding Z axis) represents the fitness of these phenotypes. The most fit genotype within this environment is indicated. Note that other peaks represent local optimal fitness, and stabilizing selection will still be active on these peaks. © Michael DeGregori

While we have been discussing the similarities in the evolutionary forces of mutation, selection, and drift that control organismal and somatic evolution, there are relevant differences. For starters, somatic cells do not have sex, and thus the sexual selection so important for organismal evolution is absent in the soma. Within organismal populations, sex can have a stabilizing effect on phenotype, thought to result from the choice of mates with minimal apparent changes and through gene flow in an interbreeding population, which limits speciation. For this reason, the isolation of a sexually reproducing population, often requiring geographic separation, is thought to frequently drive speciation (Mayr 1982). No such barriers exist for somatic cells, which behave essentially as an asexual population. Moreover, somatic cells lack the ability for sexual recombination that is so useful for generating adaptive genetic combinations, as well as for eliminating alleles that reduce fitness.

Purifying selection is weaker in somatic tissues, since a given cell type can dispense with many genes that are not required for its specialized function (Yadav, DeGregori, and De 2016). For example, genes specifically required for neuron function will be dispensable for most other cells, and thus somatic mutations in these genes will not affect the fitness of these nonneuronal cells. Thus, neuron genes will be under very weak purifying selection in nonneuronal cells, and mutations will accumulate in these genes. In addition, all somatic cells possess chromosomes from each parent and thus have two alleles of most genes, except for those on sex chromosomes. This diploidy can buffer a cell's phenotype from change resulting from the mutation of one of two copies of a gene. Either the cell can maintain its phenotype with only one copy expressed, or dosage compensation—whereby the expression of the unmutated copy is increased to compensate for the loss of its sister copy—can temper the effects of mutation. Purifying selection may also be weaker in somatic tissues, given that some stem cell pools are small and/or fragmented, and stem cells typically divide infrequently. All of these factors will reduce the ability of purifying selection to eliminate stem cells with mildly deleterious mutations. Purifying selection may be even weaker for cancer cells, which have dispensed with the need to contribute to normal specialized tissue function. For example, some leukemias arise from blood stem cells. These cells would normally need to maintain intact genes for the generation of diverse types of blood cells, from antibody-producing B-cells to red blood cells. The resulting leukemias lack these constraints, and in fact losing the ability to differentiate into many blood lineages could be advantageous. Thus, somatic purifying selection might be very weak, and many mutations could accumulate in otherwise important genes.

A final important distinction between germline and somatic evolution is that the latter is self-limited, as cancer almost always dies with its host (with notable exceptions; see Ujvari, Gatenby, and Thomas 2016). Thus, somatic evolution will be limited by time and cell generation number in its ability to generate new adaptations and innovations. Accordingly, as pointed out by Frédéric Thomas and colleagues (2015), cancer cells should not be subject to kin selection, whereby the recognition of related individuals drives cooperative or altruistic behavior. Moreover, there will not be selection for cancers that are less pathogenic, and that allow better coexistence with the host, as there is for certain viruses and bacteria. In this sense, it is too bad that cancers lack foresight, as otherwise they would realize that their destructive behavior was going to bring about their own demise as well.

What Evolution Is

Given the focus of this book on the evolutionary biology of cancer, it is important to consider what evolution is not: progressive, goal oriented, benevolent, or malevolent. And evolution cannot predict the future. These facts are important but frequently misunderstood. For instance, humans are not the inevitable consequence of 3.5 billion years of evolution. To focus on one class of animals, had a giant meteor not struck the earth off the Yucatan Peninsula sixty-six million years ago, or had there not been other changes late in and after the Cretaceous period (including continental drift and changing sea levels), it is doubtful that the earth would be inhabited by the same mammal species that are present today. The environment would have been very different, thus selecting for different types. Evolution does not necessarily lead to greater complexity or intelligence—in fact, the most dominant life-forms on earth are still microbial.

We need to think of cancer in this same light: it is not progressive or goal oriented. Instead, cancer evolution represents a series of somatic adaptations to changes in the cancer's tissue microenvironment—very much like species evolution. Selective pressures from the microenvironment can favor the evolution of either a more malignant or a more benign cancer. Of course, chance is also critical, in terms of the mutations that occur and whether they persist. An altered environment creates opportunities for adaptation, and some cancer cells need to have the right genotype and phenotype to take advantage of this opportunity. We must always return to the three major forces of evolution (mutation, drift, and selection) and the context-dependence of their action as we seek to understand who develops cancer, why some people do and others do not, when it occurs, and how it evolves within us.

To understand cancer evolution within us, we must also understand how species have evolved on our planet. We must appreciate how dramatic changes in the Earth's atmosphere, land masses, ocean chemistry, temperatures, and biota have dramatically affected evolutionary trajectories. Microbes, plants, and animals have adapted to new environments, evolving strange and innovative forms and functions that never cease to amaze. We must also understand the converse: that stable environments promote the status quo in organismal populations. It is hard to improve on what is well adapted. I will continue to draw parallels between the Earth's biota and its changing landscape, on the one hand, and the evolutionary forces that control somatic evolution in our tissues, on the other hand, to build the argument that classical evolutionary principles are paramount in somatic evolution, which is driven by changing tissue landscapes and stymied in healthy tissues.

CHAPTER 3

The Evolution of Multicellularity and Tumor Suppression

FOR ITS FIRST 2-3 BILLION YEARS, starting around 3.5 billion years ago, life on earth was unicellular. The first eukaryote, with organelles like mitochondria and nuclei, is thought to have arisen about 2.5 billion years ago. Almost another 2 billion years would pass before the first simple animals (metazoa) came on the scene. Metazoa are composed of different groups of eukaryotic cells that perform different functions for the benefit of the entire organism. Multicellularity likely provided protection from predators and could also facilitate predation or otherwise improve food acquisition. However, it came with a cost, in that only a fraction of the cells (the germ cells) contributed to future generations, with intergenerational disposal of the soma. Multicellularity also engendered a new risk: that evolution at the somatic level would produce rogue cellular expansions that could impair the fitness of the organism. In this chapter, I will discuss the mechanisms that have facilitated the evolution of animals, while preventing the somatic evolution and cancer that would otherwise have countered any fitness gains derived from multicellularity and large size.

Multicellularity and the Problem of Cancer

As described in Chapter 1, the evolution of sexually reproducing animals created the soma, which from a purely biological perspective is simply a vehicle for the germline. For single-celled organisms, there is no somatic or germline cell lineage: the organisms themselves, even sexually reproducing ones, contain the genetic information that is passed on to the next generation. Since cancer is a disease of uncontrolled growth of somatic cells, I will primarily discuss the evolution of somatic cells, and how tissue microenvironments influence this evolution. Nonetheless, it will be important to consider evolution at the germline level: how animal evolution and the environmental pressures that drive evolutionary change have affected cancer incidence and the resulting mortality.

As noted above, a true multicellular organism is made up of different cell types that perform different functions for the benefit of the whole organism, with many cells not directly contributing to the next generation. These somatic cells essentially have one ultimate goal: to maximize the potential for the germline cells to pass their genetic material on to future generations. Complex multicellular organisms such as vertebrates consist of multiple functionally specialized cell types organized into tissues, including striated muscle and intestinal epithelium. Tissues, in turn, are organized into more complex structures called organs, with multiple tissue types coordinating to perform specific bodily functions such as digestion, respiration, reproduction, and decision making.

The slime mold *Dictyostelium* may provide clues about the pressures that led to multicellularity (Du et al. 2015). These eukaryotes normally live as unicellular organisms that feed on bacteria in the soil. When food becomes scarce, starvation stimulates *Dictyostelium* to form relatively complex multicellular structures of up to 100,000 cells, with a stalk and a fruiting body that releases spores into the air. There can even be a motile phase, during which this multicellular community moves as a "slug" toward light, presumably to better position the fruiting body for the release of spores. Notably, the *Dictyostelium* cells that form the stalk do not contribute to the released spores and thus constitute a sort of soma (note the hedging—we cannot really call this structure an "animal"). The cells in the stalk increase the odds of reproductive success for the released spores, but they themselves will not contribute to the dispersed next generation. Still, the released spores are likely to be highly related to the cells that form the stalk, so from the perspective of kin selection, the trait of contributing to the stalk increases fitness. The strategy of forming the stalk and fruiting body evolved to maximize the continuity of the genetic lineage during food scarcity.

Theodosius Dobzhansky (1937) described evolution as a change in allele frequencies in a population over time. These changes can have functional consequences for the organism, and thus selection will act on them. We can similarly consider somatic evolution to be changes in allele frequencies in cells of the soma. Throughout this book, I will refer to genetic alleles, since different alleles are the basis of different phenotypes. One class of genetic alleles that I will discuss frequently consists of genes with oncogenic mutations, which contribute to the cancer phenotype. Some somatic evolution is beneficial for the organism, such as the evolved mechanisms of clonal selection that occur in our immune system. However, as a whole, somatic evolution is disruptive and can even be lethal. Thus, natural selection has worked to limit it.

Genetic changes that create new alleles can affect somatic cell fitness. Similar to the classical definition of fitness, somatic cell fitness is the average propensity of a cell of a particular genotype to increase or decrease in frequency in a cell population (Rozhok and DeGregori 2015). If this cell population is the stem cell pool, and if a genotype increases the odds that a stem cell will be lost from this pool, then this genotype would be said to reduce fitness within this pool. In contrast, a genotype that provides an advantage to the cell, increasing somatic fitness, will lead to the expansion of cells with this genotype in the population. The evolution of animals immediately created a problem: how does the animal avoid the excessive expansion of some somatic cell lineages, which would compromise the ability of the whole animal to survive and reproduce? As I discussed in Chapter 1, somatic evolution that can lead to cancer or other disruptive malignancies is not typically a major impediment to the reproductive success of the organism.

Cellular Societies in Tissues

Scientists have offered contrasting explanations for how somatic evolution is limited in multicellular organisms. The most common explanation is that the evolution of multicellularity involved the surrendering of cell fitness (Fortunato et al. 2016). In line with this view, somatic cells lost their "self," and their selfishness, to contribute to the maintenance and proper functioning of the organism. Accordingly, the reacquisition of selfishness, through mutations that increase somatic cell fitness, is thought to contribute to the development of cancers.

I propose an alternative explanation: within a multicellular organism, single cells maintain their competitiveness and high cellular fitness as a way to limit somatic evolution by limiting the frequency of adaptive mutations. This explanation posits that the very characteristics that allowed single-celled organisms to succeed (high cellular fitness, and thus competitiveness) would also serve the tissue as a whole. High fitness of cells within a tissue can facilitate the elimination of damaged and even potentially cancerous cell clones. Of course, natural selection has also led to certain penalties imposed on a cell that goes rogue, including cell death—so that there are limits to how selfish a cell can be.

Cancer as Somatic Evolution: The Enemy Within

It should now be apparent that somatic evolution follows many of the same rules as organismal evolution and is acted upon by many of the same forces. Our bodies, and those of all other animals, are essentially disposable. Animal evolution will act to limit somatic evolution and avoid cancer to maximize the odds of successful germline transmission to the next generation. Anticancer strategies are essential components of overall animal fitness, but only to the extent that they occur when reproduction is still likely. When cancer reduces your chance of reproducing, there will be strong selection against genotypes that confer risk of such cancer.

Cancer is much more than just a somatic cell clone that divides wildly and out of control. The evolution of cancer involves the acquisition of many new traits that increase the aggressiveness of the cancer, allowing it to invade its host tissue and spread to other tissues, a process referred to as metastasis. Mutant clones that simply divide into a mass but do not spread are referred to as benign tumors, and these can typically either be ignored (like many of the moles on our skin) or be removed surgically.

Cancer is distinguished from other forms of somatic evolution by its aggressive and damaging behavior. The word "cancer" is derived from the Greek word for crab, as early Greek scholars recognized the crab-like invasion of cancers into tissues. It is this metastatic spread of cancer to other organs, not the primary tumor, that frequently leads to the death of the patient. Leukemias and lymphomas, which are cancers of the blood-producing (hematopoietic) system, are mobile at all stages and infiltrate all parts of the body.

In two of the most widely read papers in all of cancer biology, Douglas Hanahan and Robert Weinberg described a set of hallmarks of cancer (Hanahan and Weinberg 2011 and 2000). These hallmarks include dysregulated pro-growth signaling, unrestricted proliferation, avoidance of cell death (immortality), attraction of blood vessels, invasion of neighboring tissues and metastatic spread to other organs, altered metabolic control, and increased mutation rates. Cancers avoid immune destruction and are often associated with increased inflammation. Typically mediated by particular oncogenic mutations, these hallmarks contribute to the cancer phenotype. Despite this, cancer evolution is not simply mediated by the time-dependent accumulation of the oncogenic mutations required to engender the full set of cancer phenotypes. Instead, the acquisition of these phenotypes results from selection for mutations adaptive to the substantial hurdles that the evolving tumor encounters. Robert Gatenby and Robert Gillies (2008) described how the growing tumor creates its own hurdles. As one example, when a tumor grows beyond the size where diffusion from nearby blood vessels can provide it with nutrients and oxygen, it typically must evolve the capability to attract blood vessels into itself.

The ultimate genotype and phenotype of a cancer is highly dependent on the selective barriers that shaped its evolution, as these barriers determine which mutations are positively selected. It is important to consider that these barriers themselves evolved at the organismal level, as part of a tumor suppressive tool kit that limits cancer through the reproductive years. The occurrence of cancer is tragic, but for every cancer that evolves to the point of being life-threatening, there were likely thousands of cell clones that made it part way down the road to being a cancer but failed to clear some hurdle. Tumor suppression usually wins.

Mutation Rates and Evolvability

As described above, mutations provide the phenotypic variability upon which selection can act. Natural selection has acted to limit mutation rates, since most mutations and phenotypic changes will reduce organismal fitness. However, the ability to limit mutation rates has been shown to be dependent on the effective population size of the organism (Lynch 2010). The effective population size can be described simply as the number of individuals in a population who on average contribute genetically to subsequent generations—a number that is often smaller than the actual population size. For small organisms like bacteria or yeast, this number can be very large (many millions), and thus natural selection has been effective at selecting for DNA replication and repair machinery that ensures a very low mutation rate (on the order of one mutation per every 10^{10} to 10^{11} bases per generation). In contrast, for large organisms, such as elephants and whales, population sizes are much smaller, and thus germline mutation rates are higher. While there are currently more than seven billion humans on earth, for most of human history it is estimated that effective population sizes were on the order of ten thousand people, based on modern genetic variation (Eller, Hawks, and Relethford 2004). Ancient human populations likely experienced local population extinctions that led to frequent genetic bottlenecks.

Effective population sizes shape germline mutation rates, as larger population sizes increase the ability of purifying selection to eliminate genetic alleles that increase mutation rate (for example, by encoding less efficient DNA polymerases), as higher mutation rates often reduce fitness. Of relevance to our discussion of cancer, somatic mutation rates are likely constrained by the evolved replication and repair machinery that limits germline mutations. It follows that the evolution of tumor suppression required for the evolution of large and long-lived animals, including humans, did not involve reductions in the occurrence of mutations in the soma.

Despite the facts that evolution has largely kept mutation rates low, and that large animals do not appear to have limited cancers by reducing mutation rates, increasing mutation rates does help accelerate evolutionary change. Mutator phenotypes in bacteria provide a great example (Galhardo, Hastings, and Rosenberg 2007). When subjected to stress, bacteria can switch from high-fidelity polymerases to mutation-prone ones. While the low-mutation phenotype is the most advantageous under stable conditions, as the organismal phenotype is already well adapted, a mutator phenotype can be advantageous in an unstable or changing environment. The higher mutation rates provide greater odds that some new phenotype resulting from a new mutation will be adaptive in the new environment. Most mutations will still negatively affect fitness, but survival of at least some of the genetically similar bacteria in the new adverse conditions may necessitate a substantial phenotypic shift.

In fact, many cancers have been shown to have higher mutation rates than the normal tissue from which they were derived (Hanahan and Weinberg 2011). As noted in Chapter 2, the increases in mutations can be in the form of point mutations in the code, additions or deletions of genetic material, and large-scale rearrangements of chromosomes. In a minority of cases, this genetic instability results from the familial inheritance of gene alleles that reduce DNA repair or maintenance. As an example, a rare syndrome in humans is hereditary nonpolyposis colorectal cancer (HNPCC), also known as Lynch syndrome. The affected individuals are at greatly increased risk for the development of colon cancer. Back in the early 1990s, investigators noticed that these cancers were characterized by certain types of DNA mutations (with extra bases inserted or lost) that are typical of defects in a known DNA repair pathway. The defective pathway (called the mismatch repair [MMR] pathway) had been extensively characterized in bacteria and yeast. The investigators applied their understanding of the genes involved in the MMR pathway in bacteria and yeast to humans and showed that homologs of the same genes operating in these unicellular organisms are responsible for MMR defects in humans (Müller and Fishel 2002). Several of these genes were shown to be mutationally inactivated in individuals with HNPCC. These individuals inherit one bad copy of an MMR gene, and the second allele is lost during cancer evolution.

Having a higher mutation rate (a mutator phenotype) provides the evolving cancer with more genetic diversity upon which selection can act, increasing the odds that the malignant clone will evolve to overcome each of the barriers that arise.

Other cancers can somatically acquire mutations that confer a mutator phenotype, thus helping fuel somatic evolution. These mutations can inactivate key repair genes (such as the MMR genes), as well as genes important for detecting the DNA damage and orchestrating a response. An example of the latter is the p53 gene, which is a transcription factor that turns on and off many genes in response to DNA damage or other cell stressors. P53 is referred to as the guardian of the genome, given its central role in mounting responses to DNA damage to protect the integrity of the genome (Vousden and Lane 2007). Not coincidentally, the p53 gene is lost in about half of all cancers, and the pathway that controls the gene is disrupted in many of the other cancers. Cells lacking this gene have higher rates of genome instability, including chromosomal rearrangements.

At this point, readers may notice a contradiction. I have discussed how changes in genetic material are almost always disadvantageous to the organism or cell. Just because a mutation that leads to a mutator phenotype might benefit a tumor's evolution at some point in the future does not overcome the fact that such a phenotype will be disadvantageous when it occurs. In the present, other cells will likely outcompete the cell that just acquired the mutator phenotype, as this phenotype typically reduces cellular fitness. The mutator cell clone will likely be extinguished, with loss of the potential for the increased mutation rate to generate cells with adaptive mutations. Nonetheless, mutator phenotypes do arise somatically. This is seemingly a paradox, but there are two possible explanations.

First, the mutation causing the mutator phenotype could also result in other phenotypes, and it is one or more of these other phenotypes that are acted upon by selection. For example, the loss of p53 activity confers resistance to hypoxia, which results from the poor vasculature present in cancers (Hammond and Giaccia 2005). P53 gene disruption by mutation would be selected for due to this hypoxia, since selection would be acting on the phenotype of hypoxia resistance. Still, we know that p53 loss also confers a mutator phenotype. In this case, the mutator phenotype may come along for the ride, hitchhiking with the hypoxia resistance. The cost of the mutator phenotype in terms of the more frequent occurrence of disadvantageous mutations is compensated for by the hypoxia resistance phenotype, which provides an advantage that is greater than the disadvantage. Note that the compensation should occur only in particular contexts, such as in the hypoxic tumor microenvironment. Should the same p53 mutation occur in a cell that is in a healthy, well-oxygenated tissue, the costs would outweigh the benefits, and the mutation would be unlikely to persist.

Second, given that the ability of selection to eliminate a disadvantageous mutation becomes weaker as the population size decreases, mutations conferring mutator phenotypes will be better tolerated in small somatic cell pools, particularly when the mutator phenotype does not drastically reduce cellular fitness. For example, colon cancers exhibit somatic loss of MMR genes, in the absence of the inherited disrupted allele (Boland and Goel 2010). Stem cells in the colon are organized into small separated pools of 10–20 stem cells that are sequestered in indentations of the intestinal lining called crypts. A mutation would need to have a significant impact on cellular fitness (increasing it by roughly 5–10 percent) to overcome the drift barrier—whereby chance would be dominant over selection in determining allele frequencies in these cells. Consequently, a mutation that disrupts MMR genes (assuming that it does not overly reduce fitness) may persist long enough in one of these crypts (and even numerically expand by drift in the pool) to increase the probability that a subsequent mutation will occur that confers an advantageous phenotype. Such a mutation could compensate for the fitness cost of the mutator phenotype. This reasoning could in part explain why mutations that result in mutator phenotypes are more common in cancers arising in epithelial tissues with stem cells organized into many small pools, and substantially less common in leukemias and lymphomas that initiate within a single large and intermixing stem cell pool in the bone marrow (Thomas et al. 2017; Kandoth et al. 2013).

In fact, for MMR gene mutations, both explanations may be correct. MMR genes are important not only for repairing DNA mismatches, but also for detecting and alerting the cell to the presence of these mismatches

(Toft et al. 1999). For many cells, this alert system can activate a particular process of cell suicide known as apoptosis. The cell commits suicide for the good of the tissue and organism, rather than potentially propagating the mutations that could result from mismatches—that is, it is better for the cell to be dead than dangerous from the animal's perspective. Certain dietary carcinogens (including natural ones) cause DNA lesions that are repaired by MMR genes, and selection for deficiency in these genes has been shown to occur during exposure to certain chemotherapeutics. Exposure to enough of such a carcinogen (whether artificial or natural) could therefore lead to selection for MMR mutations in intestinal stem cells, because not repairing the DNA mismatches may be better for the stem cell than recognizing the damage and signaling for apoptosis. In this case, from the stem cell's perspective, it is better to be dangerous than dead. Again, the advantage for acquiring a deficiency in MMR genes would be highly context dependent and would likely require changes in the intestinal microenvironment away from the microenvironment shaped over evolutionary time, whether caused by synthetic chemicals or due to the good fortune of reaching old age. In concert, the trait of changing more (such as the somatic adaptability conferred by a mutator phenotype) would increase the odds of successful navigation of changing fitness landscapes, as long as the mutator phenotype can initially be tolerated either by hitchhiking with another phenotype or by drift-mediated persistence in a small stem cell pool.

Whether amplified by mutator phenotypes or not, somatic evolution has the potential to interfere with an animal's reproductive success. Germline evolution has thus acted to limit somatic evolution. Given the inevitability of mutations with each division of a somatic cell, how can we and other animals be so good at avoiding cancer through much of our lives?

CHAPTER 4

Mechanisms of Tumor Suppression

T HE EVOLUTION OF ANIMALS—PARTICULARLY those with large bodies and long lives, like vertebrates—necessitated avoiding tissue-destructive somatic evolution and cancer. Tumor suppression is a key component of organismal strategies to maximize reproductive success. Cancer is not a simple disease, and the mechanisms that animals have evolved to prevent cancer and its destructive consequences are many and varied. This chapter will break these mechanisms down into three categories: cell intrinsic mechanisms, such as cell suicide programs, to prevent the propagation of cells with potentially damaging mutations or other disruptions; cell extrinsic mechanisms, such as immunity, that work from outside the potentially dangerous cell to limit oncogenic progression; and mechanisms integral to the evolved organismal system, such as hierarchical tissue organization.

Cell Intrinsic Mechanisms

Cells of animals—even ones that, like sponges, have extremely simple tissue structures—have evolved mechanisms to maintain appropriate numbers of cells within tissues. Cell division and survival are regulated by various social cues, including growth-factor stimulation, contact with other cells, attachment to an extracellular matrix, and adequate blood supply. These control systems exhibit substantial redundancy, with feedback mechanisms that penalize rogue cells. Somatic cells in animals have evolved numerous ways to minimize the propagation of cells that acquire genetic mutations or other damage, or experience the consequences of mutational activation of an oncogene or inactivation of a tumor suppressor gene (Lowe, Cepero, and Evan 2004). These evolved mechanisms are critical to maintaining the proper functioning of a tissue, as damaged cells are likely to be less useful and thus should be eliminated from the tissue. These mechanisms clearly also provide potent cancer suppression, by purging tissues of potentially oncogenic cells.

One key way to achieve these ends is quite simply to prevent mutations or other damage from accumulating by detecting and repairing any damage. All organisms, whether single-celled bacteria or complex mammals, possess myriad different DNA repair mechanisms that detect and correct mismatched bases (such as a C paired with an A instead of a G), broken DNA strands (both single- and double-stranded breaks), base insertions or deletions (the addition or removal of a base), and damage to bases (such as from ultraviolet light, oxidative damage, or chemical attack) (Lombard et al. 2005; Blanpain et al. 2011). Importantly, these systems are quite effective, and mutation rates are very low: a couple of dozen mutations or less with each duplication of a mammalian somatic cell genome—which is impressive when one considers that billions of bases are replicated in a single cell of a vertebrate animal. Other systems also detect and repair damage to organelles like the endoplasmic reticulum and mitochondria, or recycle damaged proteins and lipids in our cells (Green and Levine 2014). Stem cells appear to have one additional trick to avoid damage—expressing high levels of pumps located in their outer membranes that expel potential toxins, including carcinogens (Goodell et al. 1996). These pumps are energetically expensive, which could explain why their activity is highest in the stem cells that must be maintained for a lifetime. All of these mechanisms maximize functioning while minimizing the potential to propagate damage.

Despite effective repair systems, given the size of our genome and the trillions of cells in our body, DNA mutations happen in cells every day. Most of these mutations have minimal impact on cell phenotype or function and are typically tolerated. Thus, our cells will accumulate changes in our genomes through development and throughout our lives. Nonetheless, our cells utilize highly conserved mechanisms to eliminate the clonal propagation of damaged cells (Lowe, Cepero, and Evan 2004). First, DNA damage can be detected and signaling pathways activated, which leads to the death of the cell via suicide. One of the best studied of these cellular suicide processes is apoptosis, from the Greek word for "falling off." Just as trees have programmed mechanisms to lose their leaves during winter months, multicellular organisms have evolved mechanisms like apoptosis to eliminate cells. This process is important for sculpting tissues during development, selecting a potent but safe immune cell repertoire, eliminating cells with function-disrupting damage (damage not just to DNA, but to other cellular constituents as well), and purging tissues of potentially oncogenic cells.

Cell suicide in response to DNA damage can be independent of the phenotypic effects of the induced mutations. If a cell experiences DNA or cell structure damage that exceeds some threshold, then the cell institutes an apoptotic program, thus eliminating itself from the tissue. This threshold will vary for different cell types. For example, some structural cells with low chances of oncogenic transformation (because of their low potential for cell divisions) appear less susceptible to damage-induced apoptosis, compared to highly replicative cells, such as those in intestines (Blanpain et al. 2011). Evolution has favored this apoptotic strategy where the costs of nonaction are the greatest. For highly replicative cell types, it is better for the cell to die than to risk propagating an acquired oncogenic mutation. Nonetheless, this threshold cannot be so low such that cells are constantly eliminated in a tissue with each cell division, which is inevitably accompanied by some mutations. Evolution tuned this threshold to maximize tissue function while minimizing the risk of transformation.

In addition, cells can recognize phenotypic changes that could lead a cell down the path of oncogenic transformation. Mutations that either create activated oncogenes (typically by enhancing or deregulating their normal functions) or that inactivate tumor suppressor genes (preventing them from carrying out their normal functions, such as growth suppression) might be thought to have set the wheels of cancer evolution in motion, but millions of years of evolution have associated the re-sulting phenotypic changes to cellular reactions leading to elimination of the mutated cell. For example, the MYC gene normally plays critical roles in organismal development and tissue maintenance. Mutations that increase MYC expression, such as amplification of the genomic locus that contains MYC (leading to many copies of the MYC gene), will lead to hyperactivation of MYC. MYC is a transcription factor, regu-lating the expression of a large number of genes involved in cell growth and division (Pelengaris, Khan, and Evan 2002). Thus, gene amplifica-tion greatly boosts the expression of these genes.

Fortunately, evolution has linked the activation of MYC to the trig-gering of apoptosis, including through the activation of p53 (Lowe, Cepero, and Evan 2004). In the right context, such as a tissue cell that needs to divide to promote tissue growth or maintenance, the activation of p53 and other apoptotic responses are counteracted. However, if MYC activation is either too strong or in the wrong context, the cell will often be eliminated through a p53-dependent pathway. Similar apoptotic pathways are activated in response to loss of the retinoblastoma tumor suppressor gene (RB), and this apoptosis can also require an intact p53 pathway. In experimental models, elimination of p53 can prevent the apoptosis that results from MYC activation or RB loss from occurring. Notably, cancers that do have MYC amplification or RB genetic inacti-vation often bear inactivating mutations in p53 (Nilsson and Cleveland 2003; Karachaliou et al. 2016). The evolution of multicellular animals required the acquisition of new tumor suppressor gene functions that limit inappropriate cell expansion, and it is not surprising that the origin of metazoans coincides with the emergence of a number of oncogene and tumor suppressor founder genes (Domazet-Lošo and Tautz 2010).

Apoptosis is just one mechanism for eliminating damaged or poten-tially transformed cells. There are other cell death mechanisms. Some of these are more akin to execution, as other cells can recognize damaged or poorly functioning cells and then engulf and digest them. In addi-tion, cells that experience damage will undergo a process called cellular senescence, which is a permanent exit from the cell cycle (Lowe, Cepero, and Evan 2004). Cellular senescence is the ultimate consequence of cel-lular aging, observed in cell cultures after many cell divisions as well as in tissues of older humans. Senescent cells can no longer divide, but they remain metabolically active. As with apoptosis, this response can vary depending on cell type, amount of damage, and context. For example, following exposure to radiation (which damages DNA and other cellular constituents), fibroblasts—a type of tissue cell that helps create a structural matrix—will undergo senescence (Coppé et al. 2010). In contrast, the progenitor cells that maintain our blood will undergo apoptosis (Blanpain et al. 2011). These differing cellular decisions have been determined by natural selection and thus are presumably the most beneficial for the organism. Perhaps inducing senescence for a fibroblast, which has minimal transformation potential, allows the cell to continue to perform its function in situ, with less disruption of tissue function. For other cell types, damage or oncogenic pathway activation can promote cell differentiation toward a more mature state (Fleenor, Higa, et al. 2015). This process will lead to the elimination of the damaged or mutated cell clone, as differentiated cells typically lack the potential to divide and are short-lived.

In the 1970s and 1980s, some of the earliest studies to use newly discovered oncogenes showed that expression of various oncogenes in cell culture was sufficient to lead to transformed (cancer-like) phenotypes, contributing to the common view that cancer progression is limited by the occurrence of the corresponding mutations. A simple conception of mutations in cancer emerged: after the mutation of an oncogene, the gas pedal is always pressed down, while after the mutation of a tumor suppressor gene, the brake is disabled. To this day, oncogene and tumor suppressor gene functions are frequently studied in cell culture models, removed from the normal tissue context. Moreover, the cells used are typically immortalized and/or partially transformed, often lacking p53 function. In the 1980s researchers in the laboratories of H. Earl Ruley and Robert Weinberg demonstrated that when oncogenes (mutant versions of RAS genes, including HRAS, KRAS, and NRAS) were introduced into fibroblast cells that had been freshly isolated from rodent embryos, the oncogenes failed to induce transformation. Transformation by mutant RAS genes required either the expression of a second oncogene or the loss of a tumor suppressor gene (Land, Parada, and Weinberg 1983; Ruley 1983). In fact, RAS expression alone impaired the expansion of the cell clone. This concept of oncogene cooperation contributed to a better understanding of multistage carcinogenesis. Nevertheless, if the first oncogenic event impairs cell proliferation, what are the odds that the next event will occur?

Just as MYC activation or RB inactivation can induce apoptosis, the activation of other oncogenes or inactivation of tumor suppressors can lead to senescence. Indeed, Manuel Serrano, Scott Lowe, and colleagues showed that mutational activation of RAS genes causes cellular senescence, indicating that this mechanism was not limited to aging (Serrano et al. 1997). Since senescence represents a permanent withdrawal of the cell from the replicative pool, and most mutations occur during cell replication, the chances of further malignant progression are greatly reduced. A mutation, such as in p53, that prevented the oncogenically initiated cell from undergoing apoptosis or senescence might be able to allow progression. However, by thwarting the clonal expansion of the oncogenically initiated cell, the probability that further oncogenic mutations will accumulate within the initiated clone is very low. The math is simple. If we estimate that the chance of generating the next mutation along the path of malignant progression is one in a million (10^{-6}), then the odds of this occurring in a single cell is one in a million. In contrast, without these apoptotic and senescence safeguards, the clone could expand, multiplying this probability. If the clone size surpasses a million cells, then this probability of the next oncogenic mutation starts approaching certainty. Since the phenotype of an advanced cancer requires the accumulation of multiple mutations in the same clone, preventing clonal expansions following each potentially oncogenic mutation is a powerful evolved strategy to limit cancer risk.

Finally, there is another mechanism that limits the excessive expansion of cells—the maintenance of chromosomal ends (Blasco 2007). Each chromosome of a eukaryotic cell is essentially one long double-stranded DNA molecule, tightly wrapped and organized with protein handlers. The replication of eukaryotic chromosomes creates a problem, as DNA polymerase requires a primer for initiation, which is supplied in the form of RNA. The DNA landing pad for this RNA primer is not replicated at the ends of chromosomes, and thus chromosomes should shorten with each round of DNA replication. If this shortening reaches a critical degree, then a cell undergoes crisis and can activate DNA damage responses that lead to apoptosis or senescence. All eukaryotes have an enzyme (telomerase) that can add repetitive sequences to the ends of chromosomes, known as telomeres. For a single-cell organism like yeast, telomerase is always active, and thus telomeres do not shorten with each cell division. The same applies to germline cells of vertebrates and other animals, to allow for continuous passage of genetic material across many generations. For somatic cells, however, telomerase activity can be limited, which may be particularly relevant for long-lived stem cells.

While the requirement for telomere maintenance is certainly not limited to multicellular organisms, progressive telomere shortening in somatic cells would be expected to limit tumor development. An oncogenically initiated clone that overcame hurdles like apoptosis and senescence would still be limited in its ability to divide indefinitely by the crisis created by shortening telomeres. All of these hurdles evolved to limit the probability that oncogenically initiated cells can evolve to the point of being dangerous to the host organism.

Cell Extrinsic Mechanisms

Animals have evolved multiple mechanisms to eliminate potentially oncogenic cells or cancerous growths that require the action of noncancer cells. These mechanisms include cell competition and immunity.

In the 1970s, Gines Morata and colleagues described a new phenomenon in the fruit fly *Drosophila melanogaster*: cell competition (Morata and Ripoll 1975; N. Baker 2011). They generated mutations in developing wings of the fruit flies that tempered cell function, such as by reducing ribosomal proteins or the levels of MYC. While these cells were capable of forming the whole wing and indeed a whole fly when all cells were mutated, when these cells were made to compete for space and resources with wild-type cells, they were rapidly outcompeted. (Given genetic variability in populations, there is no true "wild-type" cell or organism—I refer to wild-type cells as those lacking mutations that result in phenotypic change away from the population norm.) The mutated cells with reduced somatic cell fitness are eliminated by apoptosis, and they can even be engulfed and digested by their wild-type neighbors.

Importantly, these phenomena are not limited to flies but are also evident in mice, where they use the same signaling pathways (Claveria et al. 2013). Thus, these mechanisms were likely already present in the last common ancestor of insects and vertebrates over half a billion years ago. These processes appear to be critical for maintaining healthy tissues, eliminating cells with suboptimal fitness as a quality control mechanism.

These same mechanisms can eradicate cells with oncogenic mutations. While it is often assumed that oncogenic events increase cell fitness, this is actually not true in the majority of circumstances. In fact, oncogenic mutations in fly cells can lead to the active elimination of the cell by its wild-type neighbors (Ballesteros-Arias, Saavedra, and Morata 2014; Menéndez et al. 2010). Well-conserved signaling pathways in both the wild-type cells and to-be-eliminated oncogene-bearing cells orchestrate the purging of the latter. A similar mechanism has also been shown to remove oncogenically mutated epithelial cells from a monolayer, which is a tissue organization scheme found in much of the aerodigestive tract and skin. Here, the mutated cell is forced out of the monolayer by its wild-type neighbors (Kajita and Fujita 2015). This eviction from the neighborhood leads to a nasty demise for the oncogenically initiated cell—by extrusion into the lumen for intestinal cells or shedding into the outer world for skin cells.

Vertebrates have evolved complex immune systems to eliminate foreign pathogens. These systems can also destroy malignant cells. The immune system can be conceptually divided into innate and adaptive components, although we know that the two work hand in hand. Both invertebrates and vertebrates possess innate immune systems. These systems can detect foreign invaders using pattern-recognizing receptors on immune cells such as macrophages and neutrophils (Janeway and Medzhitov 2002). These and other white blood cells can engulf and destroy bacteria and other pathogens, secrete toxic hydrogen peroxide, and even release nets of sticky DNA to entrap pathogens. The pattern-recognizing receptors that these white blood cells bear can identify certain types of molecules common to whole groups of pathogens, such as the lipopolysaccharide that constitutes the membranes of many bacteria. In vertebrates, following recognition of invading pathogens, innate immune cells signal to the adaptive immune cells that danger is present. Adaptive immune cells include B- and T-lymphocytes (known simply as B- and T-cells). The former make antibodies that recognize and destroy pathogens, and the latter can recognize virus-infected cells.

In addition to limiting pathogen infections, these systems can target malignant cells for destruction, thus contributing to tumor suppression. While antibodies made by B-cells can directly recognize foreign molecules (antigens), T-cells recognize only small pieces of proteins (peptides) displayed on cell surface proteins called major histocompatibility complexes (MHC). One type of MHC displays peptides from pathogens engulfed by cells like macrophages, alerting T-cells to the presence of these invaders. Of relevance for cancer prevention, a second type of MHC displays peptides from most of the proteins expressed within any given cell. Since T-cells have been educated during their development to recognize only "non-self" (that is, foreign) proteins, when our normal cells display peptides from normal proteins, T-cells do not react (except for in individuals with autoimmunity). However, if a precancer or cancer cell displays a peptide from a mutated protein, T-cells may be able to react with it. T-cells that react with a target cell can either kill that cell or help mount an attack by other immune cells. Just presenting a mutated peptide will not be enough, as other danger signals are needed (Iwasaki and Medzhitov 2015). These other signals could come in the form of tissue destruction or other perturbations mediated by the growing tumor, or the resulting inflammation—a potent danger signal. Mutations happen all the time in our cells, and thus the immune system has evolved to be selective in its killing ways.

While animals have evolved potent immune systems to eliminate both invaders and cancers, cancers are evolving somatically, and successful ones will overcome immune barriers. Full-blown cancers have typically evolved the trick of avoiding immune attack, and in fact they often cajole the immune system into being their minions. For example, macrophages that should be orchestrating attacks become essential components of the tumor microenvironment, suppressing T-cells, secreting factors that support tumor growth, promoting blood vessel growth into the tumor, and even remodeling the matrix to facilitate invasion (Condeelis and Pollard 2006). Up to half of the mass of a tumor can be immune cells, but these immune cells have switched sides and essentially are enslaved by the tumor. Fortunately, decades of research into the mechanisms of this immune suppression have led to the design of strategies to overcome it. These strategies typically involve reawakening of the suppressed T-cells by blocking the T-cell suppressive proteins expressed on macrophages, tumor cells, and the T-cells themselves (Khalil et al. 2016). These strategies are leading to what appear to be cures for some patients with advanced, and previously incurable, cancers.

Mice and humans with compromised immune systems are at increased risk for certain cancers (Mortaz et al. 2016). For example, individuals suffering from acquired immunodeficiency syndrome (AIDS) are at increased risk for certain lymphomas, sarcomas, and other cancers (Boshoff and Weiss 2002). Interestingly, eliminating B- and T-cells in mice increases cancer risk, but only for a few normally rare cancers. There are clearly backup immune mechanisms. Another type of white blood cell, called natural killer (NK) cells, also appears to play a key role in eliminating cancers, as mice with engineered mutations that prevent the formation of NK cells are at substantially increased risk for cancers (Waldhauer and Steinle 2008). We can imagine that one way an evolving cancer might avoid immune detection is to turn off the expression of its MHCs. Without the presentation of peptides to the constantly probing T-cells, these tumor cells attempt to go incognito. Fortunately, NK cells recognize cells that fail to present this identification card and, as their name suggests, kill these cells. Another important component of NK cell recognition of cancer cells or virally infected cells is the recognition of NK ligands, proteins expressed by these cells that signal "something is wrong with me" (other proteins can ward off NK cells, essentially saying "I'm okay"). Nonetheless, some tumors do still evolve reduced expression of MHC as a mechanism of immune escape, which overall must still be advantageous.

Tumors also evolve to simply not present antigenic peptides. Robert Schreiber and colleagues have called this process "elimination, equilibrium and escape" (Mittal et al. 2014). It is not as if the tumor makes a choice to get rid of proteins with mutations that might be recognized by T-cells. Rather, somatic evolution is again at work. Tumor cells bearing peptides recognized as foreign by T-cells are simply eliminated by the T-cells, and thus there is strong selection for tumor cells that lack these immunogenic proteins (selection that can lead to tumor cell escape from T-cell mediated killing). In some cases, the cancer cells and the host immune system reach an equilibrium, where tumor cell replication is offset by cell elimination by immune cells, and in such cases tumors can remain dormant for years, if not for an entire lifetime.

Our immune system works to purge malignant cells, even though a tumor can still sometimes win this somatic evolutionary battle. This conflict is reminiscent of the evolutionary battle between hosts and parasites. For example, worms are endemic in the large intestines of many humans living in poverty in much of the world, and we evolved with these worms over thousands if not millions of years (Jackson et al. 2009). To avoid elimination, the worms evolved mechanisms to suppress our immune systems, and humans in turn evolved a stronger immune system to better control the worms, which led to the evolution of worms with even better immunosuppressive abilities, and so forth.

Humans cannot evolve nearly as fast as pathogens—particularly viruses and bacteria, which can undergo many more generations in a day than humans can over a thousand years. Likewise, cancer cells can divide daily, and at least the successful ones will evolve mechanisms to avoid immune destruction, even though this immune destruction is multilayered and involving innate immune, NK, and T-cells. Therefore, malignant somatic evolution would appear to outpace our ability to mount defenses. This is where the adaptive immune system comes in, as each immune response represents somatic evolution in action, selecting for a limited number of B- and T-cells that can respond to a threat, whether pathogen or tumor. This responsive system complements the array of other tumor suppressive mechanisms that we have evolved to limit cancer, some of which are common to all vertebrates (like B- and T-cell responses) or most animals (like tumor suppressor genes and apoptosis).

Mechanisms Integral to the Evolved Organismal System

Tissue organization may also contribute to the "peer pressure" exerted on malignant cells, whereby a normal microenvironment can suppress the malignant phenotype. Early experiments by Beatrice Mintz and colleagues showed that injection of embryonal cancer cells into a mouse embryo at an early stage of development resulted in these normally cancerous cells' contributing to many normal tissues in the resulting mice (Mintz and Illmensee 1975; Stewart and Mintz 1981). When injected into an adult mouse, the same cells formed disorganized and aggressive cancers. The converse is also true, as injection of normal embryonic cells into the "wrong place" (such as under the kidney capsule) led to carcinoma development, leading the authors to propose "a non-mutational basis for transformation to malignancy and of reversal to normalcy" (Mintz and Illmensee 1975, 1). Being in the wrong neighborhood can lead to bad behavior, independent of genotype.

Others have proposed a field theory of carcinogenesis, according to which carcinogens primarily function by disrupting dynamic interactions among cells and structures in a tissue (Soto and Sonnenschein 2004). Harry Rubin (2007) has described how the development of a complex organism, with many distinct cell types working in an orchestrated fashion, favors cellular conformity. Conversely, the disruption of tissue architecture in old age causes a loss of this cellular peer pressure, contributing to oncogenesis. Similarly, experiments demonstrate that mutagens can promote carcinogenesis independent of the induction of oncogenic mutations. Rita Cha, William Thilly, and Helmut Zarbl (1994) showed that carcinomas induced by the DNA damaging carcinogen NMU (N-nitrosomethylurea) arise from HRAS mutations that existed before NMU exposure. Elegant studies by Maricel Maffini, Carlos Sommenschein, and colleagues showed that stromal cells (like fibroblasts and other structural cells), not the epithelial cells from which the carcinoma originates, are the targets of NMU action in a mouse model of breast cancer (Maffini et al. 2004). NMU treatment of mice prior to transplantation of unexposed epithelial cells sufficed to promote cancer development. Similarly, earlier studies by Mary Helen Barcellos-Hoff and Shraddha Ravani (2000) showed that irradiation of the mammary fat pad promoted carcinogenesis from transplanted unirradiated (oncogenically initiated) mammary epithelial cells. These studies challenge the conventional view that carcinogens cause cancers by inducing oncogenic mutations. Instead, the studies support the idea that microenvironmental disruption plays a role in fostering carcinogenesis. Finally, elegant experiments by Mina Bissell, Valerie Weaver, and colleagues further demonstrated how cell contact with the surrounding matrix, and its stiffness and architecture, can dictate cell behavior, even converting malignant cancer cells into well-behaving contributors to a tissue (Bissell and Hines 2011; Pickup, Mouw, and Weaver 2014). Clearly, context (in this case, the microenvironment) matters.

Above I discussed how cell context can determine whether a cell that receives a growth promoting signal moves forward with division or commits suicide. So how does a cell know what the right context is? We and other animals have evolved tissues with complex interactions and requirements. Cell replication is necessary for the growth of tissues during development and the maintenance of tissues in adults. Even in adults, tissue damage will evoke repair and regrowth, and immune cells can expand in numbers rapidly and dramatically in response to an infection. All of this cell expansion and tissue homeostasis requires careful orchestration of responses. For example, during homeostasis, tissue cells should divide only when new cells are needed. In response to a wound, cell divisions occur to the point that the tissue structure is restored, and no more. Even with an immune response, generating billions of lymphocytes, the vast majority of these immune cells must die once the infection is cleared to reset the system to prepare for the next invasion. These decisions are primarily dictated by extracellular cues—signals coming from outside of the cells that are dividing or dying. For a cell to divide or survive, it has to receive these external signals, including via contact with the matrix and soluble growth and survival factors. In fact, often the default program for a cell is to commit suicide, so that it must be told to live (Thompson et al. 2005). One can quickly appreciate how such a complex set of requirements for cell survival and proliferation can be tumor suppressive, as a cell that might gain an oncogenic mutation and thus disobey extracellular cues would be forced to commit suicide or become senescent. Thus, the problem is solved (most of the time).

A highly underappreciated but profound concept advanced by a number of investigators, starting with John Cairns, is that the evolution of animals has been constrained by the requirement to avoid somatic cell evolution that disrupts normal tissue function (Cairns 1975; Leroi, Koufopanou, and Burt 2003; Crespi and Summers 2005). These evolutionary constraints on tissue development and structure should be more critical for larger and longer-lived animals, and those that require tissue renewal during adulthood. The hierarchical organization of tissues, with a small number of stem cells at the top of the hierarchy followed by progressively more differentiated cells, likely plays a significant role in effective tumor suppression. While stem cells for most tissues must be maintained throughout life, the much more numerous differentiated cells are often short-lived, thus reducing the risk of oncogenic transformation.

Stem cells in our bodies are hierarchically organized, typically with a small number of stem cells generating progressively more lineage-committed progenitors that eventually make the mature fully differentiated cells that carry out the functions of our bodies. For example, the hematopoietic system of a full-grown human has 11,000–300,000 hematopoietic stem cells (HSC), which maintain the production of blood cells for an individual's entire life (Abkowitz et al. 2002; Wang, Doedens, and Dick 1997). These blood cells include the red blood cells that carry oxygen; the platelets responsible for wound healing; and the white blood cells (leukocytes) that mediate immunity, among other functions. That may seem like a lot of HSCs, until we consider that the hematopoietic system churns out more than 2×10^{11} blood cells per day (mostly red blood cells). That's 200,000,000,000 cells per day! And all of these cells are derived from this small pool of HSCs, which also divides very infrequently (about once per year). While this generation of many from few may seem impossible, a beautifully evolved system is at work. The few HSCs, by their infrequent divisions, generate a somewhat larger pool of short-term stem cells, which divide more frequently and in turn generate an even larger pool of even more committed cells termed progenitors. This process of amplification and further restriction of lineage commitment allows a greater and greater amplification of the production capacity. These progenitors are often committed to differentiate into particular lineages of cells: for example, one progenitor might make only red blood cells. Importantly, these more committed progenitors are much more numerous and divide much more frequently than the stem cells, allowing the production of massive numbers of mature blood cells necessary for survival.

As we have learned, cell divisions generate risk, in that mutations occur with each division, and some of these mutations could contribute to cancer. Herein lies the beauty of the system: Progenitors, which are numerous and divide often, are short-lived. They will undergo a huge burst in proliferation to make a large number of mature cells, but they themselves will eventually differentiate into the same mature cells. Thus, even if a potentially oncogenic mutation occurs in one of these progenitors, the most likely fate of this mutation will be to end up in fully differentiated cells, thus minimizing the potential for contributing to a cancer. Essentially, this process represents tumor suppression by differentiation.

Fully differentiated cells are often short-lived. In the hematopoietic system, neutrophils (a key white blood cell important for clearing bacteria) only live 6–8 hours in circulation and are engulfed by other cells after their death (along with any potential oncogenic mutation that they may have acquired). Analogously, for the skin and large intestine, differentiated cells are essentially on a conveyor belt, being shed every week or so into the outside world. Thus, if a progenitor cell in the intestine acquires an oncogenic mutation, in most cases this cell (and its progeny) will differentiate and be shed into the lumen of the bowel (or even actively forced out by wild-type neighbors, as described above in this chapter). Hence, the oncogenic mutation makes a one-way trip out the colon: tumor suppression by defecation.

In contrast, the true stem cells at the base of this hierarchy could accumulate oncogenic mutations, as they can renew themselves for the life of the organism. The small number of stem cells greatly reduces the odds of an oncogenic mutation even occurring. On top of this, given that these stem cells divide infrequently, these odds are even lower. Still, the stem cell pool cannot be too small. Natural selection had to keep the pool size for HSCs small enough to limit the target size for potentially oncogenic mutations, but big enough to optimally perform the cells' function of maintaining blood cell production for a lifetime. While we do not know that these constraints are the reasons for the very conserved size of the adult HSC pool in humans and other animals, the pool size is clearly governed by some form of stabilizing selection, whereby either a bigger or smaller pool would lower the fitness of the individual. Natural selection worked to optimize a different stem cell organization for epithelial tissues, a concept that I will return to in Chapter 10.

Finally, at an even higher level, animals have evolved instincts and behaviors that should decrease cancer risk. For example, humans and other animals are repelled by noxious smells, which can indicate toxic and/or carcinogenic compounds (Vittecoq et al. 2015). Our senses of sight and taste can also limit our exposures to carcinogenic compounds, like those contained in certain molds. However, we must remember that natural selection has acted to maximize reproductive success, not longevity, and thus some evolved tastes and desires (like our craving for protein-rich foods) serve the former better than the latter. Many humans today are also living under conditions where types of food (high in fats, sugar, and salt) that were difficult to obtain for most of the evolutionary history of our species are now plentiful, leading to a conflict between our evolved tastes and long-term health. Such mismatches between modern environments and our evolutionary history are further discussed in Chapters 12 and 13.

Undercover Tumor Suppression

The cellular and organismal programs discussed in this chapter highlight an important concept that is often overlooked: when we study cancers, we are studying the successful ones, not the many more malignant clones that started down the path to cancer but whose progress was thwarted by effective evolved tumor suppressive mechanisms (whether intrinsic, extrinsic, or integral). When ecologists and evolutionary biologists study the collection of species currently on earth, they recognize that many species came before these extant ones but became extinct. Even the fossil record is mostly a record of the winners, species that were successful enough both in terms of numbers and time to leave a discoverable fossil record. Just as species evolution is characterized by many twigs on the evolutionary bush that represent short-lived experiments in natural selection that failed relatively quickly, many cells start down the path to malignancy (for example, by acquiring an oncogenic mutation) but never reach the point of threatening the life of their host. The evolved tumor suppressive mechanisms covered in this book should provide insights into how this happens.

I have discussed a number of mechanisms that animals have evolved to avoid cancer, with the goal of maximizing reproductive success. A final mechanism of integral tumor suppression will be the subject of much of the rest of this book: the maintenance of tissue landscapes through youth that disfavor oncogenic mutations. This mechanism hinges on the ability of the environment to moderate the effects of mutations on fitness, determining the strength and direction of selection that acts on mutations. Ecology is the study of how organisms interact with their environment. Following chapters will explore how somatic cells interact with their tissue microenvironment, and the consequences for the functioning and health of the individual. In youth, the tissue microenvironment typically favors the evolved phenotype, and thus stabilizing selection will dominate. But in old age or following carcinogenic exposures, altered and degraded tissue conditions favor positive selection for mutations that are adaptive in those contexts. The strength of this selection should be dependent on the status of the tissue: the healthier the tissue, the more stabilizing the selection; the more degraded the tissue, the more positive the selection. Before we consider how tissue microenvironments determine cancer risk, it is important to discuss the current paradigm used to explain cancer causation.

CHAPTER 5

Dominant Views of the Mechanisms of Oncogenesis

THE TWENTIETH CENTURY WITNESSED tremendous advances in our understanding of cell biology and genetics in general, as well as of the underpinnings of cancer. A number of these advances were key to clarifying how cancer develops and functions. We learned that DNA is the genetic material of cells and organisms, how DNA is replicated with each cell division, and that mutations in genetic material can change cellular and organismal phenotypes. Early studies also showed that radiation exposures caused mutations in DNA and greatly increased the risk of cancer. The discovery of oncogenes as mutated versions of cellular genes and tumor suppressor genes as cellular genes that become inactivated in cancers was absolutely critical for understanding what makes a cancer a cancer (Hanahan and Weinberg 2011). Only by understanding how these mutations contribute to the hallmarks of cancer could we begin to more precisely target these diseases. Indeed, a whole new line of targeted therapies—drugs that specifically inhibit mutated oncogenes—developed over the past couple of decades. These drugs typically have fewer side effects compared to conventional therapies. A more targeted therapy should kill tumor cells at lower doses than those that would kill normal cells, creating a larger therapeutic window—the separation between drug doses that eliminate cancer and those that harm normal tissues. In contrast, many conventional therapies, such as the drugs that damage DNA, have narrow therapeutic windows, and patients can be taken to the brink of death to reduce cancer burden.

The Impact of the Oncogene Revolution

An example of how basic cancer research led to an improved anticancer therapy is the story of chronic myeloid leukemia (CML) (Deininger, Buchdunger, and Druker 2005). Before 2001, CML was an almost invariably fatal disease, and the only cures were achieved by bone marrow transplantation. Since this procedure required that the patient receive high doses of radiation and chemotherapy, which kills both normal blood progenitors and leukemia cells, followed by transplantation with donor bone marrow, the treatment of CML was associated with high rates of mortality and was not available for older patients. Unfortunately, the majority of CML patients are older. Working in Philadelphia, Peter Nowell and David Hungerford (1960) demonstrated that all patients with CML had the same chromosomal abnormality: a smaller-than-normal chromosome not present in unaffected individuals that they called the Philadelphia chromosome. Janet Rowley (1973) demonstrated that the Philadelphia chromosome resulted from a translocation between chromosomes 22 and 9, creating two new hybrid chromosomes—a discovery that identified the genetic mechanism underlying CML. Other investigators subsequently identified the fusion gene created by the translocation, called BCR-ABL. And further studies showed that BCR-ABL was a kinase—an enzyme that adds phosphate groups onto other proteins, thus changing their activity. Via its kinase activity, BCR-ABL stimulates signaling cascades to promote cell division and survival and other leukemia phenotypes. BCR-ABL was further shown to be both necessary and sufficient to cause leukemias in animal models. With this foundation, in the 1990s a young researcher named Brian Druker (2008) used a drug that could inhibit BCR-ABL's kinase activity to kill CML cells in culture dishes, then in mouse models, and ultimately in human patients in clinical trials. The results were so spectacular that the trial was ended prematurely to allow patients who had been receiving the then standard therapy to get the new drug (now called imatinib mesylate, with the trade name Gleevec). Approval by the Food and Drug Administration soon followed, and drugs that target BCR-ABL now maintain more than 80 percent of patients diagnosed with CML in long-term stable remission. The identification of the driving oncogene, BCR-ABL, led to an effective targeted treatment for an otherwise lethal cancer—a clear success emanating from the discovery and characterization of oncogenes.

Despite decades of intensive studies, accelerated by the declaration of a War on Cancer in 1971, there is still substantial debate within the cancer research community about the major causes of cancer and, equally importantly, how the mechanisms by which these causes contribute to cancer development. Some scientists argue that cancer risk can primarily be ascribed to bad luck, or the chance occurrence of oncogenic mutations that simply reflects the lifetime numbers of cell divisions for a given tissue (Tomasetti and Vogelstein 2015; Tomasetti, Li, and Vogelstein 2017). Another point of view is that most cancers are due to environmental exposures, arising from smoking, diet, pollution, and other insults (S. Wu et al. 2016). Both of these arguments are based on mathematical modeling that takes into account only one of the evolutionary forces that we have discussed—mutation. Basically, the arguments postulate that the explanation for why both intrinsic factors (cell divisions) and extrinsic factors (like smoking) contribute to cancer risk can be largely accounted for by how these factors affect the numbers of mutations. As I will argue below, both explanations are inadequate, since they fail to take into account the fundamental evolutionary concept of context-dependence of the selective (fitness) value conferred by phenotype-altering mutations.

In addition to providing new therapeutic targets, the oncogene revolution also sculpted the current paradigm for cancer causation. Since oncogenes and tumor suppressor genes are mutated versions of normal mammalian genes, and since oncogenes are able to induce cancers in animal models and are necessary for the cancer phenotype (as dramatically shown by the rapid regression of BCR-ABL-driven leukemias in patients treated with imatinib), and furthermore since carcinogenic insults induce DNA mutations, a simple paradigm was established that is still dominant today: carcinogens lead to oncogenic mutations, which in turn lead to cancer. The demonstration that cigarette smoking causes cancers, one of the most important epidemiological and public health successes of the twentieth century, further solidified this mutation-centric paradigm.

Linking Smoking to Cancer

Tobacco smoking became prevalent globally after Europeans discovered the Americas. However, not until World War I did cigarette smoking become virtually omnipresent, with each smoker using multiple cigarettes per day. In the middle of the twentieth century, the epidemiologist Richard Doll and other epidemiologists noted the sharp rise in lung cancers that paralleled this increase in tobacco consumption and used statistical methods to argue quite convincingly that smoking causes lung cancers (Proctor 2012). With strong denials from tobacco company executives and politicians, it took time before the scientists' conclusions had an impact on public policy, but the substantial reductions in smoking in the West can be traced back to their initial findings. These epidemiological studies had a huge effect on public policy, cancer incidence, and lives saved, as smoking rates in the United States have dropped from peak levels of about 65 percent of men in 1950 and about 40 percent of women in 1970 to about 20 percent for both sexes in 2010 (Siegel, Naishadham, and Jemal 2012). Lung cancer rates are dropping in parallel, with about a twenty-year delay relative to the decline in smoking. We now know that tobacco consumption is responsible for a large fraction not only of lung cancers, but also of pancreatic, head-and-neck, liver, and bladder cancers. These risks are on top of other smoking-caused disorders like chronic obstructive pulmonary disease, damage to developing fetuses, and other problems.

Starting early in the 1900s, the constituents of tobacco smoke were shown to be both mutagenic and carcinogenic (Hecht 1999). The mutagenic components were demonstrated to directly interact with and damage DNA, leading to increased mutation frequency. Mutations in oncogenes and tumor suppressor genes in human lung cancers can carry the "signature" of known carcinogens in tobacco smoke: the DNA code changes had hallmarks of smoking-induced changes, with a different spectrum from those found in cancers not associated with smoking. In terms of carcinogenicity, treatment of mice with compounds found in tobacco smoke led to substantially higher rates of cancers. These studies bolstered the epidemiological analyses that had strongly linked smoking to cancer causation. These studies have also contributed to the widespread belief that smoking causes cancer by causing mutations in DNA.

While we can substantially reduce our risk of cancer by lifestyle choices, such as by not smoking and by maintaining a healthy body weight, there is a major factor that we cannot control: aging.

The Somatic Mutation Theory of Cancer

While much of the cancer research over the past half-century has focused on carcinogens, the most dominant factor associated with cancer incidence is frequently ignored: old age. Almost 90 percent of cancers arise after a person reaches age fifty, with half of cancers diagnosed in people over seventy (Cancer Research UK n.d.). Early in the twentieth century, Theodor Bolveri (1929) first hypothesized that cancer was caused by chromosomal abnormalities. Decades later, Carl Nordling (1953) proposed that cancers develop as the result of somatic mutations occurring in cells. Peter Armitage and Richard Doll (1954) expanded on this concept, introducing the multistage theory of carcinogenesis in a highly influential paper. Based on their analysis that the age-dependent exponential increase in cancer incidence follows mathematically the sixth power of age, Armitage and Doll argued that cancer development typically requires six or seven mutations or other cell alterations. They argued that the association of higher cancer rates with old age was not due to physiological changes in the elderly but could be explained simply by the substantial time required for a single cellular clone to accumulate sufficient cancer-causing mutations. These early studies formed the foundation for the still dominant somatic mutation theory of cancer.

As mentioned above, the classic study of Nowell and Hungerford (1960) provided the first direct evidence of the genetic origins of cancer by identifying a genetic event (translocations generating the Philadelphia chromosome) strongly associated with a malignancy (CML). Nowell (1976) also developed the first evolutionary model of cancer, proposing that cancer resulted from rounds of selection for cell clones with progressively more malignant phenotypes through the sequential acquisition of advantageous mutations. Each oncogenic mutation was thought to improve cellular fitness. If so, cancer progression should be limited by the occurrence of these mutations. Moreover, as each event leads to expansion of the clone, the likelihood that further oncogenic events will occur in already oncogenically initiated cells increases proportionally to the number of the dividing cells. This pattern was believed to account for the exponential increase in cancer incidence with age.

The seminal development of this theoretical and experimental foundation for understanding cancer led to efforts to discover the underlying oncogenic events, with a huge expansion of our understanding of how cancer develops and functions, and new and more intelligent methods of treating cancer. Our current understanding of cancer development largely remains within this theoretical framework, based on the assumption that oncogenic events typically confer certain defined fitness advantages to cells relative to their normal counterparts.

However, by postulating that the occurrence of oncogenic mutations are the limiting determinants of cancer development, cancer researchers have largely ignored the critical importance of context in determining the strength and direction of selection for oncogenic events. In light of new evidence, the time is ripe for a major revision of this theory, by accounting for the substantial role of evolutionary forces beyond mutation to account for who gets cancer, and why, when, and how.

Challenging the Mutation-Centric Paradigm

The somatic mutation theory holds that oncogenic events generally improve cellular fitness, leading to their clonal selection. According to this paradigm, the occurrence of oncogenic events is believed to be the main time-limiting step in cancer initiation and progression. While widely believed, the paradigm does not fit multiple observations, including the following: about half of mutations accumulate in most tissues by maturity, oncogenic mutations are frequently detected in healthy tissues, the evolution of greater size and complexity of animals has not been accompanied by reductions in mutation rates, oncogenic events in young healthy tissues have been shown to frequently reduce stem cell fitness, and current models ignore a well-known tenet of evolutionary biology— that the fitness effects of most mutations are very context-dependent, largely dictated by the environment (DeGregori 2012). I will explore these paradigm-opposing observations one by one.

The highest rates of mutation accumulation are early in life. A number of reports suggest that roughly half of mutations accumulate during fetal development and growth to maturity (about two months for a mouse and about eighteen years for a human) (Frank 2010; Vijg et al. 2005) (Figure 5.1), which does not fit well with a model that directly links the accumulation of oncogenic mutations, age, and cancer. Similar kinetics

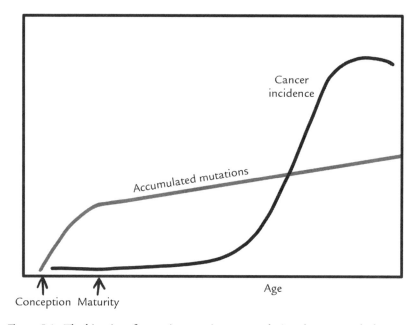

Figure 5.1. The kinetics of somatic mutation accumulation do not match the curve of cancer incidence. The "accumulated mutations" curve approximates data from mice for mutations in hematopoietic cells of the spleen (Vijg, J., et al., "Aging and Genome Maintenance," 2005) and data from humans for epigenetic changes in hematopoietic cells (Horvath, S., "DNA Methylation Age of Human Tissues and Cell Types," 2013). The "cancer incidence" curve is for humans in 2010 (based on data from the Surveillance, Epidemiology, and End Results Program) but is intended to reflect late-life cancers for any mammalian species (for more details, see DeGregori, J., "Evolved Tumor Suppression: Why Are We So Good at Not Getting Cancer?," 2011). © Michael DeGregori

are evident for epigenetic changes—heritable modifications to the DNA code that determine when and where genes are expressed (Horvath 2013). For all tissues examined, at least half of the epigenetic changes occurred by about eighteen years of age in humans. The reason for these patterns is simple: it takes a lot more cell divisions to make an animal, from fertilized egg to adult, than to maintain one. To provide one well-studied example, hematopoietic stem cells (HSCs) divide at least daily during mouse fetal development and up until almost three weeks after birth, and after four weeks of age they change to a much slower pattern, with about one division every 1–3 months in an adult (Bowie et al. 2006). Similar patterns can be inferred for human HSCs (Rozhok, Salstrom, and DeGregori 2014). Given that most mutations occur during cell division, it is not surprising that the rate of mutation occurrence will be much greater in the early part of life. Nonetheless, the reality is that we lack a good understanding of when and to what extent DNA mutations accumulate in normal tissues during a human's (or any animal's) lifetime, since insufficient numbers of individuals and age groups have been analyzed.

Even if half of all mutations that an individual will accumulate in a lifetime happen by maturity, it is possible that accumulating the multiple mutations required to make a full-blown cancer might still be delayed until late in life—the final mutations that complete the cancer phenotype might not happen until old age. However, even if we consider cancer development to represent simply the sequential accumulation of mutations, the odds of getting the full complement of mutations at maturity (at about eighteen years for a human) is mathematically about 50 percent of the odds of getting this set of mutations late in life (about seventy). Yet the risk of a human's developing cancer at age eighteen is a small fraction (about 1 percent) of the risk at age seventy (DeGregori 2011).

There is also a poor correlation between mutation accumulation and cancer risk across tissues. A recent analysis of mutation accumulation in individual stem cells from several human tissues (liver, small intestine, and large intestine) showed that a very similar number of mutations (about 2,500) accumulated per epithelial stem cell in a lifetime (Blokzijl et al. 2016). Yet cancer incidence in these three tissues are very different, being about five- and thirtyfold higher for large intestines relative to the liver and small intestines, respectively. Thus, despite previous claims to the contrary (Tomasetti and Vogelstein 2015), lifetime mutation accumulation in stem cells is insufficient to account for varying cancer predisposition across these sites.

Above I discussed how inherited mutations in the mismatch repair system lead to very high incidence of colon cancers. There are multiple other examples of familial cancer-predisposing disorders associated with the inheritance of genetic alleles that increase mutation rates (Hoeijmakers 2001). Indeed, escalating the frequency of mutations would be expected to increase cancer risk, as there would be more phenotypic diversity upon which selection could act. Still, it is important to consider that these inherited alleles do more than just confer increased mutation rates, since they can also reduce tissue function, increase inflammation, impair immune function, and cause other effects that would be expected to alter cancer risk (DeGregori 2012). Given that about half of mutations occur early in life, this link between mutation rates and cancer risk cannot adequately explain the delay in cancer incidence until late in life for most individuals. The late-life pattern of cancer incidence cannot be explained by mutation occurrence alone; rather, it requires understanding how selection for mutations is altered as we age.

Our bodies are loaded with oncogenic mutations. Another strong piece of evidence against the predominant view that oncogenic mutation occurrence is the limiting factor for cancer is that oncogenic mutations are quite common in our bodies. Studies have shown frequent mutations in or silencing of tumor suppressor genes in the breast tissue of otherwise healthy women, far outpacing the incidence of the breast cancers associated with these mutations (Crawford et al. 2004; Mutter et al. 2001). Similarly, the BCR-ABL translocation product discussed above can be found in the circulating blood of up to a third of adults, yet the lifetime incidence of the associated leukemia is about 0.2 percent (Matioli 2002). A recent study even showed that multiple clones containing several oncogenic mutations could be found in a single small patch of sun-exposed skin in humans (Martincorena at al. 2015). If even a fraction of such clones tended to progress to skin cancer, we would all be riddled with such cancers.

The math would argue that frequent occurrence of oncogenic mutations should be expected. Even with low mutation rates, the generation of the roughly forty trillion cells in a typical human will generate every possible oncogenic mutation many times over, along with every other possible mutation. Finally, even relatively complex tumors can be detected in tissues, and the vast majority of these early cancers never progress to a malignancy that would affect human health (Folkman and Kalluri 2004). For example, virtually all autopsied individuals who were ages 50–70 exhibit small carcinomas in their thyroid glands, yet the incidence of this cancer in this age group is 0.1 percent. Thus, the accumulation of oncogenic mutations, even to the point of forming small carcinomas, is not sufficient to lead to the development of cancers that threaten our lives.

Mutation rates do not inversely scale with body size. A key counterargument to the mutation-centric paradigm is the oft-ignored fact that the evolution of complex multicellularity was not accompanied by reductions in mutation rates, and mutation rates are not significantly different between organisms that have very different life spans. In fact, mutation rates are typically lower in bacteria and yeast than in mammals (Lynch 2010). As explained in Chapter 3, higher fidelity of DNA maintenance is selected for in larger organismal populations due to the ability of large populations to eliminate variants that reduce this maintenance—thus reducing fitness, even if slightly. Yet an association between aging and cancer is apparent in all studied mammals. To give one example, blue whales have about a thousand times more cells than humans and more than a million times more cells than mice, and the whales live at least thirty times longer than mice. Clearly, it takes many more cell divisions to make and maintain a blue whale than it does a mouse. Richard Peto first proposed this paradox: how is it that whales, humans, and mice all similarly avoid cancer until old age, and the overall risk of lifetime cancer is not all that different (Peto et al. 1975)? Most scientists have focused on potential differences in DNA repair, despite evidence to the contrary (Lynch 2010). I will discuss Peto's Paradox further in Chapter 11.

Oncogenic events often reduce somatic cell fitness. A major misconception underlying the current paradigm that the occurrence of oncogenic mutations limits the development of cancer is the belief that oncogenic mutations generally confer proliferative advantages on the affected cells. In other words, oncogenic mutations are thought to increase somatic cell fitness. Indeed, this assumption has been incorporated into multiple models of age-dependent cancer development in humans (see Bozic et al. 2010; Paterson, Nowak, and Waclaw 2016; Luebeck and Moolgavkar 2002). This assumption stands in direct conflict with what Chapter 4 showed about how oncogenic mutations can induce apoptosis or senescence, or lead to engulfment or extrusion from the monolayer by their covenant-enforcing neighbors. Moreover, many oncogenic mutations have been shown to contribute to loss of stem cell self-renewal (DeGregori 2012). Numerous oncogenic mutations have been engineered in mouse HSCs. When these mutations are induced within bone marrow HSCs of healthy young mice, the results are strikingly consistent: the ability of these stem cells to renew themselves is reduced by the induction of most (albeit possibly not all) oncogenic mutations. For example, while mutations in the IDH1 gene are common in acute myeloid leukemias of the elderly, which have been shown to initiate in HSCs, these same IDH1 mutations have been shown to reduce HSC self-renewal in mouse models (Inoue et al. 2016).

In the adult mouse (as in other mammals), the numbers of HSCs are kept relatively constant. Every time one of these stem cells divides, it can do so either symmetrically (producing two daughters that are both stem cells or both cells committed to follow the path toward differentiation) or asymmetrically (producing one stem cell and one committed cell), but on average it will produce one stem cell and one committed cell. Otherwise, the stem cell pool would either increase or decrease in size. A mutation that reduces stem cell self-renewal will make the odds of generating another stem cell less than 50 percent. Such a mutation will lead to exhaustion of this cell clonal lineage, and the speed of this exhaustion will be greater for larger reductions in self-renewal and higher rates of cell division. We can therefore conclude that if an oncogenic mutation occurs in a single stem cell, this mutant lineage will be exhausted and leukemia avoided. At least this should occur most of the time, particularly in a young healthy bone marrow microenvironment. Since researchers typically engineer mice so as to induce their chosen oncogenic mutations in most if not all of the HSCs, and with many of these mutations the mice went on to develop leukemia, the researchers focused on the expected result of creating an oncogenic mutation: a blood lineage cancer. However, we must remember that humans did not evolve to prevent the simultaneous induction of an oncogenic mutation in all stem cells for a tissue. Tumor suppressive mechanisms evolved to limit the development of cancer from individually mutated cells. Therefore, for the purposes of understanding cancer development, these mutant mice do not properly model the context-dependent somatic evolutionary forces that dictate the success or failure of cells that acquire oncogenic mutations.

Context-dependent fitness effects of mutations. At this point you may be wondering how and why do humans ever develop cancer? The ultimate answer lies in the evolutionary explanations provided in Chapter 1. Tissue maintenance wanes in old age, dependent on evolved strategies to maximize reproductive success. Tissues can also be damaged by extrinsic exposures. As the following chapters describe, these changes in the tissue microenvironment can alter selective pressures, including those for oncogenic mutations. Therefore, the ability of natural selection to prevent both aging and cancers declines in old age or following extrinsic damage.

This last point brings us back to a key concept in evolutionary biology: the fitness impact of a phenotype is not static. The fitness value of a mutation (or a genotype in general) is highly dependent on context, including the genetic background and, most importantly for our discussions, the environment. For somatic cells, this environment is the tissue, with its multitude of signaling mechanisms and molecules that coordinate cell behavior. To argue that an oncogenic mutation confers a fitness advantage on a cell simply by its occurrence is akin to arguing that the huge changes in life-forms inhabiting the earth resulted from stepwise mutation occurrence, as opposed to being driven by changing environments that selected for phenotypes adaptive to contemporary conditions. In Chapters 8 and 9, I discuss how changes in tissues, brought about by aging or other insults like smoking, lead to dramatic changes in the fitness value of particular oncogenic mutations.

Increasing Mutation Rates without Increasing Cancer Incidence

Mice have been generated with genetic changes that increase mutation rates, and some of these mouse strains exhibit increased rates of cancer.

Of interest, the lab of Larry Loeb generated mice with mutations in DNA polymerase δ, an enzyme that is key for DNA replication (Venkatesan et al. 2007). By disabling the proofreading functions of this polymerase, somatic mutations in the mice with these mutations increased by five- to thirtyfold. Interestingly, two different mutations at the same position led to two different results. The L604K mutation (resulting in leucine being replaced by lysine at position 604) in one of the two gene alleles led to an acceleration of the incidence of cancer which mirrored a shortening of the life span (with the age-dependent curves for both shifted leftward by several months; see Figure 5.2). Oddly, the L604G mutation (leucine being replaced by glycine) did not result either in an increase in cancer risk or rate or in changes in life span. Increases in mutation rates clearly do not suffice to change the age-dependent cancer incidence curve or the life-span curve. The fact that L640G mice exhibit substantially increased mutation rates without increases in cancer incidence is incompatible with the current theory that mutation occurrence limits cancer development.

So why do both of these curves shift to earlier ages for L640K mice? There is a trend toward greater increases in mutations in L640K relative to L640G mice, so perhaps mice with the former genotype surpass some threshold that suffices to impair cellular physiology. As shown above, life span and cancer incidence are linked, and the L640K mice provide an apropos example of this connection. The increase in DNA mutations in L640K mice may negatively affect stem cells, reducing their fitness directly and through alterations in supportive cells. As I will discuss in the following chapters, the physiological changes reflected in the shortened life span of L640K mice may create an altered tissue landscape that is more conducive to cancers. Interestingly, an examination of multiple strains of mice indicates that in general longer life spans are associated with delayed cancer incidence (G. Smith, Walford, and Mickey 1973). If aging is delayed, so is cancer.

The Persistent Paradigm

As pointed out by Thomas Kuhn in *The Structure of Scientific Revolutions* (1962), scientists who subscribe to different paradigms see different evi-

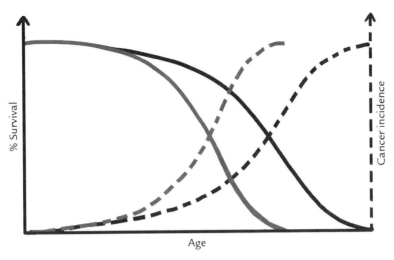

Figure 5.2. Mortality due to intrinsic physiological limitations inversely mirrors cancer incidence. Survival profiles (solid lines) of different laboratory mouse strains or breeds (black and gray) are inversely correlated with cancer incidence (dashed lines). This relationship is observed for different strains of mice as well as for mice with engineered mutations. For the latter, DNA polymerase δ L640K mutant mice slow leftward shifts in both mortality and cancer (represented in the gray curves) (Venkatesan, R., et al., "Mutation at the Polymerase Active Site of Mouse DNA Polymerase Delta Increases Genomic Instability and Accelerates Tumorigenesis," 2007). Since laboratory mice are maintained with minimal external hazards, unlimited resources (food and water), and stable environmental conditions, the survival of the mice serves as a proxy for the limits of evolved physiology or physiology altered by germline mutation. Survival is limited by the evolved programs of somatic maintenance. The waning of tissue maintenance coincides with increasing cancer susceptibility. Similar relationships are evident across species and within a species, and in humans can be affected by lifestyle. © Michael DeGregori

dence. Even with the same data, different conclusions can be drawn. Niles Eldredge and Stephen Jay Gould described how dominant theories influence how facts are interpreted: "The expectations of theory color perception to such a degree that new notions seldom arise from facts collected under the influence of old pictures of the world. New pictures must cast their influence before facts can be seen in different perspective" (1972, 83). Indeed, many scientists may simply ignore some of the facts described above. Most mathematical models of cancer incidence with age do not take into account the fact that the rate of mutation accumulation will be much greater during fetal development and growth to maturity than later in life, which presents a substantial problem for these models as formulated. This fact is not countered or denied, but simply ignored. Nonetheless, skeptical readers should obviously consider whether the paradigm currently accepted by the vast majority of cancer biologists is the correct one.

A new cancer theory is required that incorporates currently available data on mutations and on how they are acted on by selection. This theory will need to account for observations that appear to be paradoxical under the current paradigm, including the scaling of cancer incidence to evolved life spans, the effectiveness of tumor suppression independent of body size, and the critical role of environmental change in driving evolutionary change.

CHAPTER 6

Adaptive Oncogenesis

IN *THE ART OF WAR*, Sun Tzu (544–496 BCE) advised that to win battles, you must "know your enemies and know yourself" (2013). We know a lot about cancers (the "enemies"), from how various genetic and epigenetic mutations perturb signaling, metabolic, transcriptional, and genome maintenance pathways, to the genomic changes of thousands of individual cancers. Yet we do not really understand the forces that both limit and promote cancer development, nor how to fully exploit this knowledge to benefit patients. To achieve these goals, we will need to appreciate how the causes of cancer risk such as aging, smoking, obesity, and exposures to radiation and chemicals affect the key parameters of evolution—mutation, selection, and drift. As I discussed in the last chapter, previous models have focused mostly on mutation alone. In this chapter, I will introduce the theory of adaptive oncogenesis, which provides new explanations for how aging and carcinogenic exposures cause cancers. Importantly, this theory can also explain how humans and other animals avoid cancer through youth by investing in tissue maintenance. I will discuss how the causes of cancer influence the evolutionary trajectories of cells by altering tissue microenvironments, which affect selection for oncogenic mutations. In this light, and as Sun Tzu advised, we must also understand how biological systems normally function ("know yourself"), to appreciate how cancer is normally prevented during most of our lives. Knowing both the enemy and ourselves could provide the insights necessary to do a better job of preventing cancers from occurring in the first place, and to exploit cancer-specific adaptations to design more intelligent therapeutic strategies when they do occur.

Changing Landscapes Promote Evolution

While evolutionary change is often perceived to be gradual, in reality it is not. In fact, the history of life on Earth is one of dramatic bursts of speciation, coinciding with massive changes in the environment. As Peter Ward and Joe Kirschvink describe in *A New History of Life*, the physical environment of Earth has undergone spectacular changes in its atmospheric gases, most notably oxygen and carbon dioxide, and in temperature, including multiple periods of snowball Earth during which most of the planet was enveloped in ice (Ward and Kirschvink 2015). Continental drift transformed the geography of land masses and seas, changing everything from sea levels to climate. In addition, life drove many of these and other changes, such as the increases in oxygen (poisonous to early life) that accompanied the advent of photosynthesis by early bacteria, leading to the evolution of oxygen-respiring organisms.

These environmental makeovers were responsible for spectacular changes in the planet's life-forms. Each of the five major extinction events in the history of life on Earth coincided with massive changes to the environment. Environmental change resulted in maladaptation of animals and other organisms, promoting selection for genotypes adaptive to new environments. The periods between these bursts of extinctions and speciation were characterized by greater environmental and species stability. The optimization of phenotypic traits achieved with adaptation to an environment greatly limits the extent to which change can lead to further improvements.

The theory of adaptive oncogenesis draws heavily on our understanding of the evolution of species during the history of earth. Just as species evolution is driven by environmental change, somatic evolution is dictated by the state of the cellular environment. Linking cancer to its causes requires an appreciation of how these causes affect microenvironments and, consequently, somatic evolution. The cancer research community has increasingly recognized the importance of the microenvironment for cancer, although the emphasis has been on how this microenvironment positively affects cancer hallmarks like immune evasion and invasiveness. From the gene-centric perspective, the focus is often on how the genotype of the cancer affects the microenvironment, as opposed to how aging or carcinogen-altered tissue environments mold the genotype of the cancer. Adaptive oncogenesis explains how carcinogenic contexts such as aging, smoking, obesity, and radiation exposure engender cancer risk through their impact on tissue microenvironments, stem cells' adaptation to their niche, and the impact of mutations on fitness.

The adaptive oncogenesis theory has two critical principles (Rozhok and DeGregori 2015; DeGregori 2011; Marusyk and DeGregori 2008). First, natural selection at the organismal level has acted to optimize stem and progenitor cell adaptation to their tissue microenvironments. This both maximizes tissue functions and promotes stabilizing selection, eliminating phenotype-altering mutations. Of course, natural selection does not achieve perfection, and there are likely evolutionary constraints and excessive costs that limit optimal adaptation of stem cells to their tissue niches. Still, the relatively high fitness of somatic stem cells should limit oncogenesis in youth, even if not totally preventing it. Second, the functional decline of tissues brought about by aging or exposures like smoking alters tissue fitness landscapes. In nonoptimal environments, stem and progenitor cells no longer operate at peak levels of fitness, which promotes selection for mutations that can be adaptive within the altered tissue microenvironment. Some of these mutations can be oncogenic. The expansion of an oncogenically initiated clone consequently increases the risk of further malignant progression. I propose that the dominant effect of aging or carcinogens on cancer risk is to change the fitness landscape in which potentially oncogenic cells exist.

The first principle can explain why humans and other animals rarely get cancer in youth, either before or during their reproductive years. The second principle explains why humans and other animals do get cancer, providing a way to explain associations between aging and exposures like smoking, on the one hand, and increased cancer risk, on the other hand.

Why does minimizing the ability of mutation-driven phenotypic change to be adaptive limit cancer risk in youth? Moreover, why do changes to tissue microenvironments, which promote mutational adaptation, increase cancer risk? As demonstrated by Ronald Fisher (1930), the rate of evolutionary change will be proportional to population size and the mutation rate. If the mutation rate is constant, then the chance that a particular mutation will occur in a population will be dependent on the frequency of cell divisions and the size of the population. The chance of hitting a target is proportional to the size of the target. Adaptation facilitates expansion of the genotype within the population via positive selection, increasing the size of the subpopulation with this genotype and the odds of the next "hit." For example, if a mutation increases a cell's likelihood of surviving and reproducing, the fraction of cells bearing this mutation should increase in the population. In an adult, the number of stem cells and other cells in a tissue is relatively constant (homeostatic), just as most populations of animals in an unperturbed environment maintain constant numbers. Therefore, for a genotype to expand in a population at homeostasis, it needs to increase the fitness of the cells, at least for large cellular populations.

Development of a cancer represents a process of somatic evolution driven by three components: diversification of heritable types through acquisition of genetic changes, including epigenetic changes, which alter how the code is expressed; selection for cells with mutations that increase cell fitness; and random changes in genetic allele frequency through drift (Rozhok and DeGregori 2015). At least for larger cell populations, to trigger clonal expansion initiating oncogenic mutations will need to provide a fitness gain relative to other cells in the same tissue. Importantly, adaptive oncogenesis posits that the ability or inability of oncogenic mutations to cause clonal expansion is highly dependent on the tissue microenvironment, which dictates the fitness value of genetic alleles. The fitness landscape faced by a particular cell type is constituted by the phenotypes encoded within the cell population and their complex environment, including other cells, the matrix, growth factors, and nutrients. A cell near the tip of a peak on this fitness landscape is unlikely to experience a mutation that can improve its fitness, as it is already well adapted. Even if there is a higher fitness peak, representing a better adaptive solution on the landscape, reaching this other peak would be highly unlikely, requiring either a mutation of large effect to jump from peak to peak or the passage through low intermediate states of fitness, which would likely lead to extinction (Figure 6.1).

While oncogenic mutations can confer phenotypes that would appear to increase cellular fitness, such as by increasing cell division rates or promoting tissue invasion, millions of years of evolution have attached costs to these mutations that lead to a net reduction in cellular fitness. As shown in Chapter 4, the acquisition of an oncogenic mutation by a cell can lead to cell death, senescence, differentiation, engulfment by neighbors, or destruction by the immune system. While the oncogenic mutation might confer certain hallmarks of cancer, the associated costs will typically prevent oncogenic adaptation and clonal expansion,

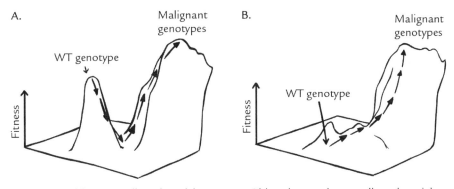

Figure 6.1. Fitness landscapes change with age or damage. As in Figure 02-01, potential genotype-encoded phenotypes are arrayed on the X-Y plane, and the protruding Z axis represents the fitness of these phenotypes. A. A hypothetical fitness landscape for stem cells in young, healthy tissue. Animals evolved stem cells to be near the top of a fitness peak, and the inherited somatic genotype is indicated as the wild-type (WT) genotype. Arrows indicate oncogenic mutations. While there is a path to malignancy, mutations that reduce the somatic fitness of cells will most often lead to the elimination of these cells by competition with more fit cells. B. A hypothetical fitness landscape for stem cells in old or damaged tissue. Damaged microenvironments with age or exposure to carcinogens engender stem cells that are no longer well adapted to their niche (even if still WT in their genotype). Since stem cell fitness is low, there is strong selection for adaptive oncogenic mutations (arrows) that can lead to malignancy. Direct damage to stem cells, such as through mutations, can also contribute to their reduced fitness. © Michael DeGregori

at least in a young, healthy tissue. In addition, by residing near the top of a fitness peak, a stem or progenitor cell in a youthful tissue will receive the "right" amount of signaling, transcription, or other activities, as determined by evolution at the germline level. More signaling via a mutated RAS protein, more MYC-dependent transcription, or less p53 transcriptional activity will actually reduce the fitness of the stem cell. Deviating from evolution-optimized levels for cellular parameters reduces adaptation to the microenvironment.

Adaptive oncogenesis proposes that long-lived multicellular organisms evolved stem cell populations with high fitness, not only as a means of efficiently maintaining a tissue, but also because high fitness in a cell population will hinder somatic evolution. Effective competition will facilitate the elimination of the occasional damaged cell from the stem cell pool, maintaining tissue fitness to maximize the reproductive success of the animal. Thus, a stem cell pool with high fitness will maintain the status quo. Maintenance of such a fitness landscape—with stem cells exhibiting near-peak fitness—requires protection of the integrity of both the stem cells and, perhaps more importantly, their tissue microenvironment. This maintenance represents an investment, likely of energy derived from food. Natural selection sculpted youthful tissue microenvironments for function that best serves the purpose of maximizing reproductive success. This tissue phenotype is called the "evolved type." The evolved type for stem cells will represent a state of high adaptation to a healthy tissue microenvironment. Thus, the stem cells in an ideal young individual will be maximally adapted to the tissue landscape. Of course, there is no ideal individual, and evolution does not reach perfection, but adaptation to a tissue through germline selection should lead stem cells to approach peak fitness.

The waning of this investment in advanced years leads to the degradation of the tissue microenvironment. The loss of tissue integrity and function underlies the physiological decline that all mammals experience, if they are fortunate to live long enough. Exposure to carcinogens, such as smoking in humans, can also promote tissue decline, and such exposures contribute to a large fraction of human cancers. In addition to increasing the frequency of mutations, these exposures can cause dramatic changes to tissues by killing cells, damaging the normally supportive microenvironment, and inducing inflammation. Carcinogen-damaged tissues present an altered environment for constituent stem cells.

Whether through aging or exposures like smoking, changes to tissues move the microenvironment away from the evolved type sculpted by natural selection. In the process, resident stem cells become substantially less fit, as they find themselves in an environment to which they are no longer well adapted. Thus, there will be positive selection for new phenotypes adaptive to this new landscape (Figure 6.1B). Some of the mutations that could improve adaptation will be oncogenic. Therefore, the same oncogenic mutation that might be negatively selected for in youth, with the result that it is purged from the stem cell pool, can be positively selected in the tissue of an old person or a smoker. This positive selection promotes the expansion of the oncogenic mutation–bearing clone, increasing the risk of further malignant progression and cancer. These changes in microenvironmental driven selection act in concert with aging and carcinogen induced alterations in mutation frequency to increase the risk of cancer development.

Cancer Risk by the Numbers

The development of cancer involves a lot of chance, from the occurrence of mutations to the role of genetic drift in populations of stem cells. Even someone who makes lifestyle choices that increase his or her cancer susceptibility, such as smoking and not exercising, could go through life without developing cancer. Equally, a young person who eats a balanced diet, does not smoke, and exercises daily can still develop cancer. Nonetheless, the risk of cancer development for the first person is far greater, just as it is for someone who is older. For this reason, we must always consider solid epidemiological data rather than anecdotes.

What determines risk at the tissue level? We have discussed fitness landscapes and how they can either favor or disfavor evolutionary change, depending on where the population sits on the landscape. These relationships hold true both for organismal and somatic cell populations. One point worth emphasizing is that natural selection, envisioned as movement on a fitness landscape, almost always proceeds one mutation

at a time: the chance that two adaptive mutations will occur before selection acts on the first one is highly improbable, given very low mutation rates. In a mammal's somatic cells, mutation rates are about 10^{-8}
(1/100,000,000) per base pair per cell division (Lynch 2010). It would
take, on average, about 10^8 cell divisions to mutate a particular base in
the genome. In large populations, each mutation must increase fitness,
as otherwise the mutation-bearing clone will likely be lost before the
next mutation could have a chance to be judged by selection. The direction of selection, by affecting the size of a cellular clone with a particular genotype, will determine the odds of further somatic evolution.

To illustrate these principles, let us first assume that for a particular
stem cell population in a particular context, such as stem cells in a
smoker's lungs, there are 100 possible adaptive mutations. Accordingly,
adaptation should on average occur within 10^6 (10^8 divided by 100) cell
divisions. If the stem cell pool size is 10^5 cells, then the chance that one
of these 100 different adaptive mutations will occur is about 10 percent
(assuming that each cell divides once). Mutations happen randomly,
providing a role for chance. We also know that cancer progression requires
multiple mutations. Estimates are typically 3–5 mutations for carcinomas,
which originate in epithelial tissues (Luebeck and Moolgavkar 2002;
Tomasetti et al. 2015). If the first mutation was adaptive in the tissue of
the smoker, the mutation will be positively selected, and the mutated cell
clone will expand in number, outcompeting its smoke-hobbled peer
cells. Let us assume that this oncogenically initiated clone expands to
10^5 cells. The chance of the next oncogenic mutation occurring following
division of all cells of the clone (again assuming that there are 100 such
adaptive oncogenic mutations) is about 10 percent. This logic applies to
two different oncogenic mutations that alter the function of two different
gene products, as well as to mutations that consecutively inactivate the
two alleles of a tumor suppressor gene. For many tumor suppressor genes,
there is selection for loss or inactivation of the second copy in a somatic
cell. Based on the reasoning above, we can understand how the loss of the
first allele must be adaptive enough to lead to clonal expansion, to increase the odds that the second copy can be inactivated.

What would be the fate of these same mutations if they did not provide a fitness advantage? Even if one of the 100 oncogenic mutations

occurred, if it was not adaptive, the cell containing it would be unlikely to expand. On average, it would remain as a single cell, with drift leading either to its elimination or its modest expansion. Thus, the chance of the next event would be roughly one in a million. If the adaptive value of a mutation is context-dependent, and the major context determining this value is the microenvironment, then changing microenvironments will greatly affect the odds of cancer development. The same math can be applied to any context that changes fitness landscapes by changing tissue microenvironments—including aging, obesity, and exposure to pollution. In each case, the odds of subsequent progression down the path to full malignancy will be highly influenced by the adaptive advantage conferred by the previous mutation in the current tissue microenvironment, with chance (mutation and drift) playing an additional role.

Considering how somatic cells navigate fitness landscapes, we can understand how natural selection at the organismal level has acted to limit somatic evolution in animals. The rate of evolution will be proportional to mutation rate, population size, and the strength and direction of selection. To limit the first of these factors, evolution could have selected for mechanisms that reduced the rate of mutation occurrence in somatic cells. However, as discussed above, the evolution of large and complex animals has not involved reductions in mutation rates. There may simply be constraints and feasibility issues that prevent further improvement in DNA sequence maintenance. Moreover, rates and extents of mutation accumulation are poorly correlated with cancer rates across tissues and species. Natural selection seems to have invested in mechanisms to eliminate oncogenically mutated cells after they are generated.

Regarding population size, the requirement to avoid cancer through reproductive years could have selected for tissue organization with low numbers of cells susceptible to cancer. Indeed, the relevant cell populations, often self-renewing stem cells, are kept small. Smaller cell pools make for smaller targets for mutations. Finally, natural selection could affect the selective advantage or disadvantage of mutations in somatic cells. A key tenet of adaptive oncogenesis is that evolution selected for animals with somatic cells that are positioned at or near local fitness optima. Cells on a local fitness peak will greatly limit somatic evolution, as evolution to a higher fitness peak would require acquisition of multiple mutations (Figure 6.1A). Progression from a fitness optimum requires progression through lower intermediate states of fitness. Within a homeostatic cell pool, a cell clone with lower fitness will be outcompeted by more fit peers. Thus, even cells acquiring oncogenic mutations would either be lost or persist at low numbers, greatly reducing the odds of the next mutation. The maintenance of stem and other cells with cancer-initiating potential at or near local fitness optima, which involves investing in the maintenance of the tissue environment, has been tuned to maximize reproductive success. In this manner, the somatic cell strategy works to optimize the fitness of the organism.

Altered Selective Pressures Drive Evolution

The predominant view in oncology is that cancer development is primarily restricted by the stepwise occurrence of somatic oncogenic mutations that confer fitness advantages on cells. This view is inconsistent with evolutionary theory, according to which the fitness value of a mutation is highly dependent on genetic and environmental contexts. Extrinsic risk factors like smoking or radiation exposure, as well as intrinsic risk factors like aging, do much more than affect mutation load: they drastically alter tissue landscapes and thus influence the selective value of mutations. These interactions hold for evolution at cellular and species levels.

The major driver of organismal evolution is environmental change, which affects selection and drift. The hominid lineage that led to modern humans has undergone drastic phenotypic change in the last five million years or so. Evolutionary biologists understand that this was not simply due to the progressive accumulation of the right mutations. Instead, changing environments and selective pressures drove human evolution. For example, brain size in hominids rapidly increased only in the last million years, at a rate much greater than the preceding four million years or so of hominid existence (Irfan 2013; Seymour, Bosiocic, and Snelling 2016). Mutations that could increase brain size likely happened many times, but increased brain size was adaptive only under the right environmental conditions, likely driven by weather patterns that gave rise to the African savannas and further promoted by the invention of fire and cooking (Wrangham and Carmody 2010). The increased calories provided through cooking are thought to have offset the increased energetic costs of a large brain. Our gorilla and chimp cousins took a different path due to different environmental pressures, not simply due to differences in mutation occurrence.

Let us consider a straightforward example of human evolutionary change involving a single gene. Lactose tolerance past the first few years of life is due to genetic alleles that allow the expression of the enzyme lactase to continue into adulthood: without those alleles, lactase expression is shut off after about three years of life. Lactase breaks lactose down into glucose and galactose, sugars that are then readily absorbed into the blood vessels of our intestines. Insufficient lactase production leads to lactose intolerance, and for individuals with that condition, the consumption of dairy products leads to nausea, diarrhea, gas, and other intestinal complications. Mutations in the regulatory regions of the lactase gene arose and rapidly expanded in different human populations at least three separate times, and only in about the past 10,000 years (Tishkoff et al. 2007). Obviously, the rapid expansion of these alleles cannot be ascribed simply to the occurrence of these mutations, which likely happened in many other populations over the millions of years of hominid evolution. Instead, the environmental conditions present during the last 10,000 years and only for specific groups of humans—including domestication of cattle and the availability of dairy products for consumption—favored individuals with these lactase gene mutations and thus promoted the expansion of these alleles in the population. Thus, this process was primarily limited by selection. In fact, alleles conferring lactose tolerance were hugely advantageous—estimates are that offspring with one of these mutations were up to 10 percent more likely to survive than those without it, simply because the former could obtain nutrients from dairy products past infancy. The fact that these mutations are not prevalent in the descendants of peoples who were not dairy farmers indicates that the mutational changes are actually disadvantageous in the absence of the availability of dairy products: millions of years of evolution had determined the "right" pattern of expression of lactase, at least in that context.

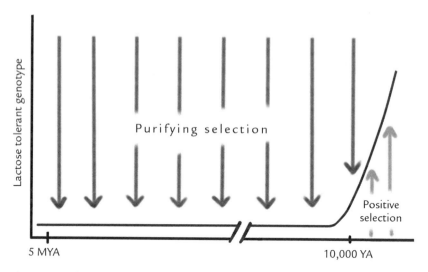

Figure 6.2. The evolution of lactose tolerance. The rapid spread of lactose tolerance alleles of the lactase gene was driven by environmental context (dairy farming), not just by mutation occurrence. The estimated frequency of the lactose tolerant genotype is depicted over the more than five million years of hominid evolution before the advent of dairy farming. For most of hominid history, the ancestral lactose-intolerant allele was favored, and DNA mutations that led to lactose tolerance were purged from the hominid population by purifying selection. In the past 10,000 years, with the availability of dairy products that provided valuable calories after weaning, the same lactose tolerance alleles were strongly positively selected. © Michael DeGregori

The evolution of lactose tolerance provides a great lesson in the different forms of selection. During the roughly five million years of hominid evolution preceding the onset of cattle domestication, the ancestral lactase gene alleles were maintained by purifying selection, eliminating mutational changes that altered this expression pattern—including the ones that later gave rise to lactose tolerance. Only in dairy-farming communities and only in the past 10,000 years did the selective pressures change for mutations providing prolonged lactase expression, which were subjected to strong positive selection (Figure 6.2). The same lactase gene mutations that had previously been disadvantageous and been purged by purifying selection, became advantageous in more recent times given the availability of dairy products from domesticated animals. In sum, the fitness impact of these lactase gene mutations was highly dependent on context.

From the discussion above, you may conclude that selection was the only driver of lactose tolerance evolution. This would be wrong. The mutations that prevent shutting the lactase gene off were required and probably also limiting. Mutations are rare, and there were likely populations of humans who domesticated cattle and who would have benefited from dairy consumption past childhood, but for whom the lactase gene mutations simply never occurred. In his thoughtful book *Chance and Necessity,* Jacques Monod (1972) described the interplay between randomly occurring mutations and the selective pressures that help determine their fate. The chance occurrence of the mutation was required, but (the necessity) it also had to occur in a population where the mutation would provide a benefit.

There is one more aspect of chance to consider: drift. Children carrying the lactose tolerance alleles were likely born in other cattle-herding communities. However, these children could have died without reproducing, or their offspring could have died without reproducing, and thus the allele did not spread in these populations. The mutation might have occurred in an individual who otherwise had been dealt a bad genetic hand, or who was exposed to a deadly infection.

Selection, mutation, and drift all contributed to the frequencies of the lactase gene alleles during hominid evolution. Nonetheless, it was selection that determined the shape of these frequencies, both in terms of timing and geography (Figure 6.2). In the case of lactose tolerance, if we took away the selective pressure resulting from cattle domestication, we would lose the pattern of evolution—the rapid increase in altered lactase allele frequency in the past 10,000 years. Thus, natural selection is responsible for the pattern of evolution, largely dictated by external conditions. If we could increase or decrease the rate of mutations, we could still not change the overall pattern, only the amplitude. The same applies to cancer.

Changing Tissue Microenvironments and Oncogenic Selection

Potentially oncogenic mutations occur relatively frequently in our bodies, but these mutations are typically disadvantageous for the recipient cells and are eliminated by purifying selection. It is only when the right tissue environmental context arises, such as in aged or damaged tissues, that these same mutations can be adaptive. It is this adaptation that allows the cell clone containing the oncogenic mutation to expand and dominate the tissue, just as people with the lactase mutation expanded and dominated populations that practiced dairy farming. This expansion then serves as the target for future mutation accumulation and adaptations.

An oncogenic mutation in a healthy lung is like the lactase mutation in a population without dairy farming: it is more likely to be lost than to persist. However, the same oncogenic mutation in a smoker's lung is more like the lactase mutation in a dairy-farming population: in this context, the oncogenic mutation can be adaptive and will quickly expand in frequency in the population (as described in Chapter 9).

Mutation occurrence is also limiting. Most smokers will not get lung cancer, even though the lung tissue landscape may be conducive to oncogenic selection. Perhaps for these smokers, the required mutation simply never occurred in cells like stem and progenitor cells that maintain replicative potential. It is even more likely that oncogenic mutations adaptive to the smoker's lung landscape did occur, leading to clonal expansions, but that further cancer progression was thwarted at a subsequent step, perhaps by the immune system or the failure to recruit blood vessels. There is also the component of drift: an oncogenic mutation could occur in a smoker's lung and in the right cell, but despite the increased odds of persisting and even expanding, this cell lineage is simply lost by chance. In this case, the bad luck for the cell is good luck for the individual.

At the organismal level, it is environmental perturbations that lead to evolutionary change, as organisms adapt to new environments. Sometimes these environmental perturbations are relatively independent of the population whose evolution is affected by these changes—for example, when changes in the atmosphere and land masses have had huge effects on the evolution of species over earth's history. In Chapters 8 and 9 I will discuss in more detail how aging or carcinogen-exposed tissues similarly dictate somatic evolution. Moreover, we can see parallels between the evolution of lactose tolerance and larger brains in humans (facilitated by dairy farming and the invention of cooking, respectively) and the way the growing tumor changes its own environment and thus dictates many of the selective hurdles that must be overcome. Just as humans' innovations can influence their own evolution, so the progression of a cancer can influence its further evolution.

Adaptive oncogenesis represents a logical extension of our understanding of organismal evolution. Selection acts on genetic diversity in a population to maximize somatic cell fitness. The fitness level conferred by a genotype is highly dependent on the microenvironmental context. Adaptive oncogenesis proposes that humans and other animals have evolved tissues whose stem cells are well adapted to their microenvironmental niches. The resulting high fitness of these stem cells leads to strong stabilizing selection, which greatly reduces the chance that a mutation can confer a fitness advantage. The well-ordered tissue state is maintained through periods of likely reproductive success, maximizing tissue function and minimizing oncogenesis. But these tissue relationships are disrupted during aging, as evolved maintenance programs wane, or as a result of carcinogenic exposures. Stem cells consequently encounter perturbed microenvironments, resulting in positive selection for adaptive mutations. Such somatic evolution can sometimes lead to cancer.

CHAPTER 7

Limiting Somatic Evolution in Youth

THE PHYSICAL AND BIOTIC ENVIRONMENTS on earth have changed dramatically over its 4.5 billion years of existence. These changes have been more episodic than gradual. While we may focus on the periods of rapid change, in between there have been long periods of stability, when both environments and the constituent species experienced more modest changes. George Gaylord Simpson (1944) was one of the first to recognize patterns of stasis in the fossil record. As I have discussed above, within a stable environment, life will evolve to be well adapted to that environment, and this state of good adaptation will disfavor change. Stabilizing selection will dominate over positive selection, as there is little room left for improvement in fitness. Environmental stability favors species stability. In this chapter, I will discuss the first key principle of adaptive oncogenesis: how maintenance of tissue microenvironments during youth promotes stabilizing selection in somatic cell pools, limiting oncogenesis and other forms of somatic evolution. I will continue to draw parallels with species stasis and organismal evolution. I will also explore some of the mechanisms that mediate stasis. Tissue maintenance is part of the overall strategy that evolved to maximize reproductive success for different species in different environments. The same maintenance strategy that maximizes tissue function and thus individual fitness during youth also limits oncogenesis, by favoring the evolved phenotype of stem and progenitor cells. Phenotypic change in these somatic cells will be disfavored. Animals have evolved a common strategy for delaying both cancer and aging: maintaining tissues and the fitness of resident stem cells.

Stasis and Phenotypic Stability

When Niles Eldredge and Stephen Jay Gould were Columbia University graduate students studying at the American Museum of Natural History in New York in the late 1960s, they studied evolutionary change in their chosen organisms, the long-extinct trilobites of the genus *Phacops* and the less ancient and even extant land snails (primarily of the genus *Poecilozonites*), respectively. Trilobites studied by Eldredge were abundant in the shallow seas for hundreds of millions of years, starting with the Cambrian explosion about 530 million years ago. The two men were struck by how hundreds of thousands of years could pass without much morphological change for many animals, occasionally punctuated by the sudden appearance of a new species. They concluded that "the norm for a species, or by extension, a community is stability" (Eldredge and Gould 1972, 115). Together they formulated the theory of punctuated equilibrium (the basis of which was first formulated by Eldredge), which proposes that instead of being gradual, evolution is characterized by long periods of species stability interrupted by occasional bouts of new speciation.

Eldredge and Gould noted that the patterns of stasis interrupted by rapid changes in species were common in the fossil record. For example, our ancestor *Homo erectus* remained relatively unchanged for about 1.5 million years. Fossils of *Homo erectus* do not exhibit gradual changes over time until specimens that resemble *Homo sapiens* or *neanderthalensis* appear. Human evolution was not gradual. This is not to say that there was no evolutionary change during periods of stasis—I have already discussed the changes in lactase gene allele frequencies in humans, changes that would obviously be missed when studying fossils. Phenotypic stability does not entail an absence of change. Nonetheless, continuity of, not change in, phenotype was dominant. The same applies to somatic cells, with minor variations in phenotype present within a tissue. Somatic evolution to generate cancer—which can be considered akin to a new somatic species—is rare.

During periods of minimal change, or stasis, natural selection is still at work. Purifying selection is particularly active in a well-adapted population, eliminating inherited genetic changes that reduce individual fitness. As Eldredge observes, "the absence of change itself was a very interesting pattern" (1999, 21). Ecology has taught us about the many interdependencies in complex ecosystems, with each species dependent on many others even in ways that might not be inherently obvious. These complex interactions contribute to stabilizing selection, as long as the environment remains relatively unperturbed. Tissue complexity, with various different cell types supporting and controlling each other through a myriad of interactions, is also stabilizing, as these complex interdependencies make it difficult for a cell to divide out of control. The cell would need to replace or supersede multiple support systems at once, such as growth factors secreted by a neighbor, physical interactions with neighboring cells that provide signals to survive, and blood vessels that provide oxygen and nutrients (DeGregori 2011). Going rogue, at least in a healthy tissue, comes with severe costs.

Almost seventy years ago, Herman Muller beautifully described how fruit flies (in the genus *Drosophila*) maintain very constant features, such as the position of their wings or the numbers of sensory bristles on the head. He noted that in the laboratory one could easily isolate mutants with alterations in traits, but that these alterations typically seemed to reduce the ability of the fly to properly function: "practically all of them do result in some quantitatively demonstrable reduction in the expectation of life" (Muller 1948, 173). In other words, the fitness of the fly was reduced. Thus, he concluded that the normal phenotype was the most advantageous, describing the "high adaptive value of precisely the 'normal' degree of gene expression now existing," and that phenotypic stability across generations is "due to active selection in favor of the normal type" (ibid.,181, 212). Millions of years of evolution has selected for optimal functions, highly adapted to the flies' environment.

As seen in Figure 6.1, in a new environment, selection will drive the phenotype of a species up the fitness landscape. If the phenotype is initially less than perfectly adaptive, then individuals acquiring mutations that increase fitness will be positively selected, moving up the peak. In contrast, individuals inheriting mutations that decrease fitness will likely be eliminated. With enough time and in a constant environment, selection will push the phenotype of the population to the top of the peak and will then act to keep it there. At the top, there is nowhere to go but down. Any phenotypic change will reduce fitness, as stabilizing selection will be strong. These selective pressures will act on every trait of the organism or cell that influences fitness. The first principle of adaptive oncogenesis is that natural selection at the organismal level has sculpted tissues that promote stabilizing selection at the somatic level. The evolved interactions between stem cells and their tissue niches will disfavor most phenotypic change in these stem cells, as long as investments are made in tissue maintenance (and in lieu of external factors such as carcinogen exposure).

As one example, Sean Morrison and colleagues have shown that hematopoietic stem cells (HSCs) exhibit an optimal level of protein translation (Signer et al. 2014). Genetically increasing or decreasing translation in HSCs led to reduced HSC fitness, as measured in competitive transplantation assays. In particular, deletion of the tumor suppressor gene PTEN increased translation but led to decreased HSC self-renewal, which as previously explained would lead to extinction of HSCs bearing this mutation. In fact, many oncogenic mutations are known to boost protein translation, and the stabilizing selection favoring the evolved levels of translation for a particular stem cell would facilitate the elimination of such oncogenically initiated stem cells.

Of course, species do change. New mutations that provide adaptive advantages may be rare, but when altered environmental conditions favor them, they have a big impact—whether as a new species or as a cancer within us.

The Evolved Phenotype

As explained in Chapter 2, purifying selection, also called negative selection, is the process whereby deleterious mutations are removed from a population, as these mutations reduce fitness. For a population that is well adapted to its environment, purifying selection will promote stabilizing selection as the current phenotype will be the favored one, and thus most mutations that change this phenotype will be eliminated. Purifying selection is strong in most organismal populations, helping maintain the status quo. This explains why some genes are so well conserved, as the requirement for their function has not changed much over long periods of evolution. Thus, an organism that inherits such a conserved gene with a mutation that affects gene function will likely be less fit, as the gene has been honed over millions of years of evolution to be just right. Nonetheless, phenotypic variability will still exist within a well-adapted population. Stabilizing selection maintains this variability within a range that is compatible with high fitness within the environment. While it might be advantageous for the population to have greater genetically determined phenotypic variability to facilitate adaptation should the environment change, excessive variance from the mean of the well-adapted phenotype will come with a cost to individuals and will thus be eventually purged from the population. This logic should hold equally well for organisms and for somatic cells.

I have discussed the evolved phenotype (what Muller referred to as the normal type), which I define as the tissue and stem cell phenotypes that maximize tissue function and stem cell adaptation, promoting good tissue functioning and stabilizing selection to limit oncogenesis. Selection for this evolved phenotype is strongest through the years of likely reproduction. There is no cliff: natural selection will not maintain full investment up to some point and then remove all investment. Instead, the investment programmed by natural selection will be strongest through development and when reproductive success is highly likely, but the investment will progressively wane as these odds become less likely. At age forty-five, a primordial human did not have a zero chance of reproduction but clearly had a lower chance than at thirty. Men are capable of reproducing into their seventh and eighth decades, but the odds of reproduction at those ages for most of our evolutionary history were very low, as few people survived that long. As mentioned above, older humans can also contribute to future generations through providing care or protection for grandchildren, and in some hunter-gatherer societies older men attain a higher social status that allows them to select younger mates

(Tuljapurkar, Puleston, and Gurven 2007). Therefore, we can see that there is no on-off switch for somatic maintenance and disease avoidance, but there are gradual reductions in the investment made that accelerate in late life in line with reduced odds of a return on the investment. We would expect the evolved phenotype to be similarly maintained during youth and then decline.

The evolved phenotype can be envisioned as what we would expect to find for a given tissue in a healthy young person. The tissue structure should be well ordered, with the proper numbers of different cell types in the proper locations. To pick just one example, in the kidney, epithelial and other cells are arranged to form complex structures called nephrons, which process waste from blood to produce urine. There are intricate arrangements of epithelial cells forming ducts and filters (glomeruli), endothelial cells forming blood vessels, neurons relaying information, and so forth. An incredibly complex orchestration of cell types, arranged in a precise fashion, allows this organ to perform its essential functions of waste elimination (while retaining useful blood components like proteins), blood pressure modulation, and general fluid homeostasis in the body. This organized and functional structure has been honed by millions of years of evolution. In lieu of an inherited disorder, our kidneys should function well through our youth by maintaining this complex and functional organization.

The waning of this investment in kidney maintenance past prime reproductive years does not mean that all of us are doomed to suffer kidney failure in old age. In fact, most of us will not. On average, the kidney of a sixty-year-old will not function as well as that of a thirty-year-old, but it will typically function well enough to prevent renal failure. In older ages, glomeruli decrease in number (with declining filtration rates), and fibrosis increases (Denic, Glassock, and Rule 2016). The numbers of kidney cysts increase. Accordingly, among the major risk factors for kidney failure is being over sixty. Importantly, the average age of diagnosis of kidney cancer is sixty-four, and this cancer is rare in those under forty-five. We can surmise that the changes to kidney structure in old age are associated with alterations in the fitness landscape for kidney stem cells. In the kidneys of a young person, these stem cells should be well adapted to their niches (of the evolved type), and thus stabilizing selection will dominate—preventing oncogenesis. However, in an older person, the same stem cells will find themselves in a microenvironment to which they are no longer well adapted, and thus they will be less fit. This maladaptation in the new environment (no longer the evolved type) will create selection for adaptive, and sometimes oncogenic, mutations. While we tend to focus on why we get kidney cancer when we are old, it is equally if not more important to consider why we do not get this cancer when young. The evolution of tissues with well-adapted stem cells was key.

It is important to stress one point: natural selection only has to be good enough to limit the probability of cancer through periods of likely reproduction. These words were chosen carefully. The risk need not be zero. As discussed above, the power of natural selection to eliminate harmful genetic alleles is reduced in small populations of animals. For example, if a particular genetic allele leads to a low cancer risk (say, one in five thousand individuals) in youth, but the effective population size is one thousand, natural selection will be relatively powerless to eliminate this genotype. Another consideration is trade-offs. For example, a genotype might confer an improved immune response through boosting the replication potential of our lymphocyte progenitor cells, but this genotype comes with a small increase in the risk of lymphocytic leukemias. We can easily envision how an improved immune response—particularly for most of our evolutionary history, when infections were a major cause of death—would increase overall fitness even with the cost of greater leukemia risk (assuming this risk did not surpass the improved survival from better immunity). Thus, the probability of cancer through periods of likely reproductive success will be limited, but not eliminated.

Similarly, I have argued above that stem cells reside on a local fitness peak and near the top of this peak. The peak is local: other phenotypes could have higher fitness in the same niche. For example, the peak for cancer could still be present in a normal youthful tissue, perhaps even attaining higher fitness than normal stem cells (Figure 6.1A). Most importantly, stabilizing selection is still at work for stem cells on a local peak, if getting from the evolved phenotype to another peak is difficult and thus improbable, requiring a rare mutation or perhaps multiple mutations. For this reason, we refer to stem cells in a youthful tissue as well adapted, not perfectly adapted.

Phenotype Stability

Based on the principles of stabilizing selection, the optimized level of expression of genes for each cell type has been selected by evolution (Muller 1948). Not only does this stabilizing selection prevent genetic change that could alter these parameters over generations, but within individuals and cells there are effective mechanisms to keep the expression and activity of different genes and their proteins well tuned. These mechanisms include feedback control of gene expression and protein stability. As an example, if in a complex more of one protein is made than can partner with the other proteins, this extra protein can be eliminated by degradation (Mueller et al. 2015). Complementing evolutionary tuning of expression at the gene sequence levels, evolution has provided further mechanisms, such as negative feedback loops, that act at the cell and tissue levels—a sort of self-tuning. Complex systems also maintain homeostasis systemically. For example, hormonal communication between the pancreas and tissues such as the liver and muscle maintain tight control over blood sugar levels (Kotas and Medzhitov 2015). Thus, in addition to stabilizing selection at the animal population level, maintaining cellular and tissue phenotypes within a narrow phenotypic range, there are also evolved mechanisms that act to buffer phenotypic change within an individual. Just as a buffered solution will resist changes in its pH despite the addition of an acid or base, there are mechanisms to maintain homeostasis in our tissues in the face of external challenges and, perhaps more importantly, stochastic fluctuations. Note, however, that of course cells respond to external cues such as damage or signals from other cells and tissues, and these responses can lead to dramatic changes in gene expression, protein levels or activity, and metabolism. But even here, tuning mechanisms will be at work, so that the responses are kept within a narrow window. The new level of expression will be tuned to a certain range. Like everything else in biology, this tuning is far from perfect, and it does not need to be perfect. There will still be stochastic variation in every parameter of a cell, whether these cells are in a resting state or responding to external cues.

Just as maintaining a garden requires work, maintaining phenotypic stability in a tissue landscape through youth comes with an energetic cost. As we know from the second law of thermodynamics, maintaining order requires energy. The DNA sequence needs to be maintained, dysfunctional proteins replaced, organelles turned over, and damaged parts renewed. Investment is thus made in perpetuating the evolved phenotype. The epigenotype also must be maintained. While the genotype is the DNA sequence (arrangements of A, T, C, and G), the epigenome consists of all the additional modifications on our chromosomes that influence how the genome is expressed (Goldberg, Allis, and Bernstein 2007). These modifications include the methylation of the cytosine base (that is, the addition of a methyl group to C) and various modifications of the histone proteins that coat DNA. These modifications dictate how genes are expressed and are critical for cell type diversification: how a muscle cell knows to function like a muscle cell, a neuron like a neuron, and so forth. Importantly, these changes to DNA and histones are somatically heritable, meaning that the epigenome is largely passed intact to daughter cells upon cell division. When a T-cell divides, it makes two T-cells. Conrad Waddington (1957) described epigenetics (before its chemical basis was known) as a landscape with hills and valleys, and a cell as being like a ball on this landscape that seeks lower ground. Different epigenetic profiles for different cell types will direct the ball into different valleys (states), leading to differentiation into various specialized cell types (Marusyk, Almendro, and Polyak 2012). All of these mechanisms to maintain phenotypic stability wane with age and are also impaired by insults like smoking. A Waddington epigenetic landscape may become less well sculpted, with a shallowing of the pockets that would normally maintain phenotypic stability—that is, a degradation of the evolved state.

The Nonlinearity of Life

Change in the physical environment on earth and its myriad life-forms has not been gradual. When we consider our tissues and bodies, change is also not linear in our lifetimes. First, the period of fetal development and growth through childhood can be clearly distinguished from the rest of our lives. This period is characterized by high rates of cell division and expansion of stem, progenitor, and mature cell pools to populate the tissues of the growing individual. Even in adults, aging is not linear. For example, we will age physiologically more from sixty to seventy than we will from twenty to thirty. This change in physiological aging is readily apparent to both observers and those experiencing old age, and it can be demonstrated by analyzing survival curves. Figure 7.1A shows survival probabilities for people of different ages in 2010. The risk of death on a per year basis, proportional to the downward slope, is flat from twenty to thirty but increases greatly from sixty to seventy. Furthermore, your risk of dying per year is 10-fold higher when you are eighty than when you are sixty (about 10 percent versus 1 percent), but only about 2.5-fold higher when you are forty than when you are twenty (0.25 percent versus 0.1 percent). After forty, aging is exponential, not linear. Alas, we pick up speed as we go downhill.

Given the relatively small fraction of individuals killed by external hazards like other humans (in murders, wars, and so on), predation (close to zero in the United States or United Kingdom), and even infectious disease (at least at younger ages), the greater risk of death in later years can largely be attributed to the inherent decline in the physiological state of individuals. The same age-dependent decline can be seen in another less grim statistic—record marathon times (Figure 7.1B). The times for both men and women are pretty much flat from about eighteen to thirty-six and then rise at an accelerating rate with greater age. Since these record times reflect the best of the best, they can serve as a proxy for the upper limits of human potential at these ages. By this measure, there is no peak but a plateau of optimal performance, which nicely coincides with expectations of reproductive success through most of human history.

While the same ten years pass going from twenty to thirty or from sixty to seventy, the impact of those years on our health and physical potential is quite different. It is therefore important to distinguish physiological aging from chronological aging. Only the former has been affected by the evolved life history strategies discussed in Chapter 1.

Given that our tissues do not change linearly but exhibit substantially less decline during youth (the years of greatest likelihood of reproduction for most of hominid history) than during older ages (when the force of natural selection to favor maintenance is weak), it follows that the cells'

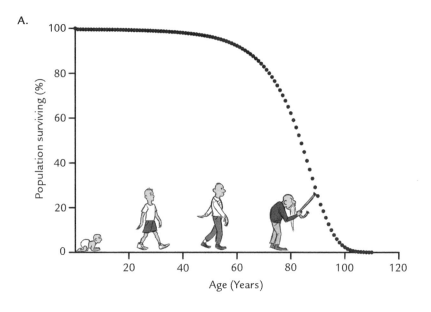

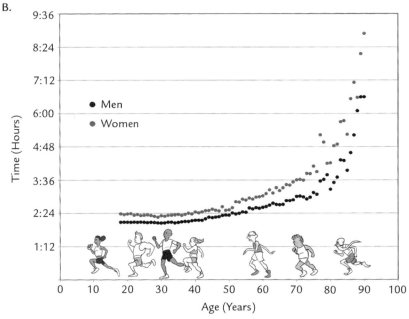

Figure 7.1. Human survival and maximal fitness potential with age. A. The percentages of survivors to each age who are assumed to be subject throughout their lives to the mortality rates experienced in 2010 in England and Wales, as representative of modern developed countries (graphed from data from the UK Office for National Statistics, https://www.ons.gov.uk). B. World records for the marathon for men (black dots) and women (gray dots) at the indicated ages (graphed from data from the Association of Road Racing Statisticians, World Single Age Records- Marathon; http://www.arrs.net/SA_Mara.htm). © Michael DeGregori

microenvironment also does not change in a linear fashion. Since the youthful tissue microenvironment favors stem cells with the evolved phenotype, stabilizing selection that opposes somatic evolution will be strong through youth. Maintenance of healthy tissues will limit selection for adaptive change, including changes that could contribute to cancers. However, in older years, the rate of tissue change accelerates. The programs that maintain tissue homeostasis are disrupted, contributing to increased disease susceptibility (Kotas and Medzhitov 2015). Just as environmental change on Earth has promoted adaptive change leading to new species, so do tissue changes in old age engender somatic evolution, including oncogenic somatic evolution.

It is less clear how evolved strategies for tissue maintenance actually lead to the curve of physiological decline. Let us consider several possible mechanisms, keeping in mind that these explanations are not mutually exclusive. First, we can imagine that a longer life strategy simply involves building a more durable body, just as we get what we pay for when we buy an appliance—the cheap ones typically break down sooner, no matter how well they are maintained. In this case, the investment is made during development. Second, the intensity of maintenance, such as how frequently cellular organelles are replaced, could be modulated. This maintenance is likely not turned off at any point, but it is eventually overwhelmed by damage accumulation. Given that certain antiaging strategies can still extend the life span when initiated in midlife, maintenance is likely a modulatable strategy (Fontana, Partridge, and Longo 2010). Third, damage accumulation could have threshold effects, whereby accumulation in youth is tolerated or buffered. Cells have two copies of every gene and other redundancies that can allow some damage accumulation without functional impairment. In addition, damaged cells in a tissue can be purged and replaced by undamaged ones. Our tissues also clearly have excess capacity (for example, a person can survive reasonably well with just one kidney). Aging of tissues may not immediately affect overall physiology. Even at the organ level, plaques can form on our arteries long before we experience a substantial increase in the risk of heart disease (Wierzbicki and Viljoen 2010). Evolution need not eliminate all damage accumulation or tissue alterations, as long as these perturbations do not significantly impair reproductive success when it matters.

Given the large increase in the risk of mortality at older ages in modern humans, we may be approaching the physiological limit of our evolved capacity to maintain the soma (Dong, Milholland, and Vijg 2016). While the average life expectancy has nearly doubled since 1900, gains in survival beyond 100 years have increased more modestly, with the maximum life span for humans remaining flat for the past 20 years (the record is 122 years). There may be limitations in the evolved structure or function of our bodies in terms of long-term maintenance. Nonetheless, perhaps a better understanding of how evolution can tune somatic maintenance to extend life spans could lead to interventions that mimic this tuning.

CHAPTER 8

Changing Adaptive Landscapes with Aging

An evolutionary biology perspective allows us to formulate new answers to fundamental questions. In this chapter, I will address the key question of why most cancers occur in old age. Adaptive oncogenesis offers an explanation quite different from current paradigms that focus on how aging affects the occurrence of mutations. I will also discuss the second principle of adaptive oncogenesis: that tissue degradation with advanced age or following carcinogenic exposures engenders selection for adaptive oncogenes. As shown in Chapter 1, the decline in tissue and organ structure and function at older ages reflects a waning of the investments in tissue maintenance, with the pattern of decline dictated by evolved life histories. I will describe experimental and computational studies demonstrating that alterations to tissue structure and function in old age change fitness landscapes. The chapter will explore the causes of tissue decline in old age and the mechanisms underlying enhanced selection for particular oncogenic events in aged stem cell pools, as shown in mice and humans. Mutations that are maladaptive in the fitness landscape of a healthy tissue can be adaptive in the tissue of an elderly individual. The risk of cancer consequently increases at older ages.

Fitness Landscapes and How They Change

In 1959, the hundredth anniversary of the publication of Charles Darwin's *The Origin of Species* (1876), it was widely accepted among evolutionary biologists that speciation was driven by the replacement of less adapted species by better adapted ones (Eldredge 1999, 1995). It was thought that more intelligent and somehow superior mammals were destined to replace dinosaurs. Thus, evolution was understood to be progressive and could be considered as limited by the occurrence of mutations that provided improvements, mirroring the current predominant thinking in cancer biology.

More recently we have learned that speciation is largely driven by environmental change, with bouts of particularly intense speciation following each of the five major extinction events known to have occurred in Earth's history. These extinction events were all instigated by major perturbations of the physical world, including continental movement via plate tectonics, changes in atmospheric gases, volcanic activity, and impacts from extraterrestrial bodies, as described by Peter Ward and Joe Kirschvink in *A New History of Life* (2015). The resulting altered environments drove strong selection for new phenotypes adaptive to the new conditions, leading to intense speciation. Evolutionary change is not limited to major earthwide perturbations. As a more recent example, plate tectonics and major climatic changes have led to punctuated changes in cichlid speciation over the last million years in Lake Malawi in Africa, leading to the adaptive radiation (diversification to fill new niches) of cichlids and the current existence of more than eight hundred species (Ivory et al. 2016; Malinsky and Salzburger 2016). As noted by Niles Eldredge in *The Pattern of Evolution*, the "almost monotonous ecosystem stability . . . is interrupted only occasionally, but inevitably (in the long run), by ecosystem-generated punctuations of environmental disturbance, followed by extinction, and, finally, speciation" (1999, 159).

As we compare somatic evolution to the evolution of species on Earth for the past 3.5 billion years, we can ask why the increased cancer risk for so many tissues follows similar age patterns. Many cancers show similar exponential increases in incidence in older ages, despite the facts that these cancers initiate in different tissues with differentially organized and sized stem cell pools and require for their evolution very different numbers of oncogenic mutations (from one to more than five). For example, chronic myeloid leukemia (CML) in its chronic phase is driven by a single oncogenic event that generates BCR-ABL, while carcinomas of the colon appear to require at least three oncogenic events and likely more (Mullighan et al. 2008; Luebeck and Moolgavker 2002; Tomasetti et al. 2015). Yet CML and colon cancers demonstrate roughly similar patterns of increased incidence at older ages, with most cases diagnosed after sixty.

The common late-life pattern, despite these differences in the associated oncogenic events, suggests a cell extrinsic causative factor. Indeed, analogous to how environmental changes on earth led to extinctions and speciations across many phyla, so general physiological decline is associated with aging's impact on multiple organ systems—which leads to selection for new cellular phenotypes adaptive to these altered tissue landscapes. While such somatic evolution need not lead to a malignancy, as adaptive mutations can be nononcogenic, some adaptive changes can be oncogenic, which increases the risk of further malignant progression. The common late-life patterns for many different cancers can thus be ascribed to their promotion via age-dependent physiological decline. The timing of increased risks for different cancers will not be perfectly synchronous. Even if conditions suitable for somatic evolution existed concurrently in multiple tissues, the resulting cancers could require different time spans to reach the point of being clinically detectable. Moreover, general physiological decline certainly will not affect different tissues with identical timing.

The magnitude of the change of an environmental factor will also affect selection for adaptive events. We can envision how a change in temperature of 1° C or a change in pH of 0.1 unit would represent a different selective force than a change of 10° C or 1.0 unit, respectively. The smaller changes might only minimally influence the fitness of the cells or organism, and a change in the genetics of the population might be evident only after many generations. Cells have mechanisms to deal with small perturbations—basically, adaptation within cells and tissues, to be distinguished from heritable changes in the genetic code that lead to adaptation that spans generations. Effective mechanisms to deal with small perturbations in the environment could dampen the selective pressure for genetic adaptation, particularly if these perturbations had no impact on survival and reproduction. The fact that the same species can be found in dissimilar environments demonstrates the existence of such mechanisms, which make the species inherently flexible. For example, coyotes occur in many different environments in the United States, and although some genetic adaptations have likely occurred, these changes have not led to the creation of new species. We can also see how human migration to markedly varied regions of the world has led to relatively minor genetic adaptations (such as for skin color), while smaller environmental changes have a minimal impact on human fitness and thus our genetics. Similarly, cells in our body are subject to certain degrees of microenvironmental variation, whether in terms of pH, glucose concentration, or the availability of other nutrients, but these changes are typically kept within tolerable ranges that cells can adapt to without genetic change.

In contrast, a large change in the environment (like a reduction in pH of 1.0 unit in a tissue) will reduce the capacity of cells to function and divide, as well as likely result in a large amount of cell death. Thus, a cell that has a mutation that increases its fitness (that is, its ability to survive and reproduce) in the more acidic microenvironment would be positively selected, with the strength of this selection proportional to its differential fitness relative to the population average. Cells with the mutation conferring resistance to acidity could still be worse off in the more acidic conditions than in the normal pH conditions. What matters is the ability of this cell clone to survive and divide relative to cells without the mutation.

Adaptation to Environmental Change

The studies of Ronald Fisher (1930), J. B. S. Haldane (1932), and Sewall Wright (1931) in the first half of the twentieth century showed that fitness can be understood as the ratio of individuals of a particular genotype from one generation to the next. If there is no change, then the fitness is 1. If there are more or less individuals of a particular genotype over generations, then the fitness is more than or less than 1, respectively. A key point is that fitness depends on context, and the context that matters most is environment. To give one extreme example, the traits of a parrot might be very well adapted to the Costa Rican forest, but the fitness value of these traits would be very low in Antarctica. Selection acts on phenotypes, and the phenotypes that influence fitness are those that are heritable—which entails a genetic (or epigenetic) basis. Thus, as a verbal summary of Fisher's equations, mutational processes generate genetic diversity, which is then acted upon by selection to increase or decrease the frequency of genetic alleles depending on their contribution to fitness in the current environment (with the additional influence of drift, which is dependent on population size).

In his geometric model, Fisher (1930) proposed a conceptual description of adaptation. Imagine a multidimensional space, in which the number of dimensions represents the number of traits. The more complex the organism, the more dimensions, with correspondingly increased difficulty in the odds of improvement through mutation. Complexity promotes stability. For simplicity, let us consider a version of Fisher's model in two dimensions—that is, with only two traits (Figure 8.1; redrawn from Orr 1998). Mutations will cause random changes in the space with respect to adaptation. If we start with a population B that is maladapted following a recent environmental change, we can see how random mutations can restore optimal adaptation. Mutations of smaller effect are more likely to be beneficial, since they are less likely to overshoot the zone (circle) of improvement, but they are also more likely to be lost by chance.

While Fisher used his model to argue for gradual evolution in very small steps, subsequent evolutionary biologists argued that the more maladapted a population is, the greater the number of mutations that can confer adaptation, and that these mutations can have a greater effect on the phenotype (and thus on fitness). Studies of adaptation in model organisms from microbes to vertebrates demonstrate that adaptation can involve mutations of large phenotypic effect. As adaptation is reached, mutations of smaller effect will be favored, with more and more phenotypic change decreasing fitness (Orr 2005). When you focus a microscope, initially you make large changes in the dial. As you start to attain focus, you turn the dial less and less. Similar stepwise adaptation may occur during cancer development, with initial mutations having greater impacts on phenotype (Zhao et al. 2016). There is evi-

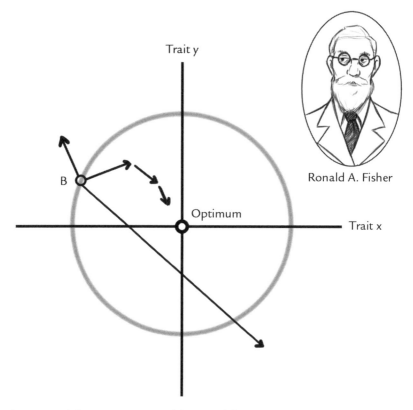

Figure 8.1. Fisher's geometric model. A simplified two-trait model is shown. After an environmental change, the average values of traits X and Y for a population B no longer confer optimal adaptation. Arrows indicate random mutations in individuals in population B. Mutation-induced phenotypic change that results in trait values closer to the optimal phenotype will be adaptive, and they include all new trait values within the circle. Mutations of large effect will be more likely to overshoot the bounds of the circle. After each adaptive mutation, the new genotype will be closer to the optimum, and mutations of smaller effect will be more likely to be adaptive. Model predictions should be relevant for both organismal and somatic cell populations. Redrawn from a similar graph in Orr, H., "The Population Genetics of Adaptation," 1998. © Michael DeGregori

dence for recurrently mutated oncogenes and tumor suppressor genes in cancers, and many of these mutations can have substantial effects on phenotype, and do appear to occur early in tumor evolution. Nonetheless, some of these driver mutations can occur late, but this may reflect that cancer cells seeking adaptation are chasing a moving target, as the changing microenvironment during cancer growth promotes selection for new adaptations.

Studies in simple organisms like yeast beautifully show how changing conditions influence evolutionary trajectories. Daniel Bolon and colleagues showed how mutations in a key gene involved in stress responses, HSP90, always reduced the fitness of yeast under standard growth conditions (Hietpas et al. 2013). As Wright noted, gene frequencies tend "to remain constant in the absence of disturbing forces" (S. Wright 1931, 105). However, a small subset of these HSP90 mutations improved fitness under harsh conditions, like elevated salinity. Thus, there is substantial potential for adaptation to a new environment even in a very well conserved gene. Interestingly, mutations that were adaptive in high salinity were less fit in the standard conditions—there was a cost to adaptation. Such a cost is perfectly predicted by Fisher's geometric model, as movement to a new optimum will logically involve movement away from the previous optimum. These experiments in yeast reinforce three basic principles that are relevant to somatic cell evolution and adaptive oncogenesis: the evolved type confers (locally) maximal fitness under the environmental conditions that the population is adapted to, changing these environmental conditions changes the fitness value of particular gene mutations (increasing the adaptiveness of some), and mutational adaptation to a new environment will decrease fitness in the old one.

Returning to the fitness landscape in Figure 6.1A, in a young healthy lung, the evolved landscape will be an ordered one, theoretically with a major peak representing the evolved state for the stem cells, dependent on the youthful tissue microenvironment. The tip of the evolved peak can be considered as analogous to the optimum in our example of Fisher's geometric model (Figure 8.1). However, natural selection did not act to develop programs to maintain this landscape in old age—nor, for example, in the lungs of smokers, since humans did not smoke for the vast majority of our evolutionary history. Thus, the landscape will be substantially altered in old age or following carcinogenic damage. In a cellular population that is poorly adapted to its environment, cells will occupy new positions on the landscape. Not only will the stem cells reach lower heights of fitness, but there could also be new peaks of potential adaptation on the landscape, engendered by altered microenvironmental conditions. There will be strong selection for mutations that improve fitness, moving a stem cell up a peak (Figure 6.1B). The more perturbed the landscape (within limits), the greater the number of mutations that could be adaptive, with mutations that would have a larger impact on the phenotype more favored.

Differential Selection for Oncogenic Mutations with Age

I have discussed how tissue changes, resulting from relaxed germline selection for tissue maintenance in older ages, promote selection for adaptive oncogenic mutations. Using fitness landscapes and classical evolutionary theory, I have also presented potential mechanisms for both stabilizing selection and oncogenic adaptation in tissues. Nevertheless, up to now support for adaptive oncogenesis has been presented as purely inferential and theoretical. A critical reader will want evidence.

The greatest prognostic factor for development of cancer is old age, and the average age of onset for many carcinomas is about seventy. The incidences of hematopoietic malignancies like CML, acute myeloid leukemia, and chronic lymphocytic leukemia also increase logarithmically in old age. According to the theory of adaptive oncogenesis, age-dependent microenvironmental changes in tissues are key to somatic evolution leading to cancer. Curtis Henry, Andriy Marusyk, and colleagues (Henry et al. 2015; Henry et al. 2010) showed that when the oncogenes BCR-ABL, NRAS, and MYC were introduced into young bone marrow progenitors in a young bone marrow environment in mice, the oncogenes did not provide a significant advantage to the progenitors, rarely leading to clonal expansion and leukemia (Figure 8.2, top panel). In contrast, when the same oncogenes were introduced into old progenitors in an old bone marrow environment of mice, rapid clonal expansion ensued, leading almost invariably to leukemia formation (Figure 8.2, middle panel). Oncogene-driven leukemia genesis was limited not by the occurrence of these mutations, but by the age of hematopoiesis. Importantly, even when the oncogenes were inserted into young cells that were then introduced into old bone marrow environments, selection for the oncogenes and enhanced leukemias occurred. Moreover, oncogene-expressing old cells did not form leukemias when placed in a young bone marrow

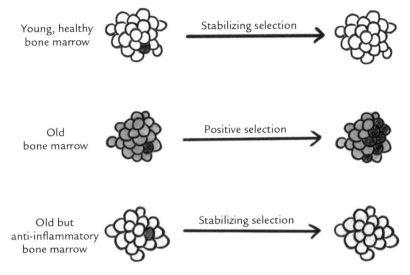

Figure 8.2. Age and inflammation promote oncogenic adaptation. Top: In a young healthy bone marrow microenvironment, stem and progenitor cells are near a fitness optimum (the lightest cells), and therefore stabilizing selection will dominate. Phenotype-changing mutations, including oncogenic ones (the darkest cells), will likely be maladaptive, and cells bearing these mutations will be eliminated from the pool. Middle: In an old bone marrow microenvironment, stem and progenitor cells will no longer be well adapted to their niches. There will be positive selection for adaptive oncogenic mutations, leading to clonal expansions and increased risk of further progression toward leukemia. Bottom: The inhibition of inflammation in old mice leads to a more youthful bone marrow microenvironment (with fitter cells, shown in speckled white), which promotes stabilizing selection for the evolved phenotype. Oncogenic mutations will be maladaptive. © Michael DeGregori

environment: the presence of young fit bone marrow progenitors suppressed leukemia genesis. Additional studies from the laboratory of Hartmut Geiger showed that an aged bone marrow microenvironment promotes clonal hematopoiesis (blood cell production) and pre-leukemic cell expansion (Vas, Senger, et al. 2012; Vas, Wandhoff, et al. 2012). The age of the microenvironment, not the age of the cells receiving the oncogene, is key.

Decades of study by numerous labs have shown that hematopoietic stem and progenitor cells in the bone marrow of old mice or humans exhibit defects in their ability to be maintained and to function (Ergen and Goodell 2010; Beerman et al. 2010; Henry, Marusyk, and DeGregori

2011). Old hematopoietic stem cells do a poor job of reconstituting hematopoiesis when transplanted into another individual, and old B-cell progenitors are poor producers of mature B-cells. In the elderly, these defects lead to increased incidence of anemia and decreased ability to mount immune responses, including to vaccines. While infectious disease is no longer a major killer of the young in the developed world, viral and bacterial diseases still cause a substantial fraction of deaths in the elderly. Studies have showed that B-cell progenitors in old mice exhibit multiple defects that would be expected to reduce cellular fitness. These defects include impaired signaling from receptors on the surface to influence cellular programs, reduced expression of genes that are key in cell growth and replication, and dramatic impairment in metabolism (such as the ability to make nucleotides, the building blocks for RNA and DNA) (Henry et al. 2015; Henry et al. 2010; Van der Put et al. 2004; Lescale et al. 2010).

Elucidation of these defects led to the development of a hypothesis about how particular oncogenes could be adaptive in old progenitor pools. Could expression of the oncogenes correct the defects? Indeed, Henry and colleagues showed that many of these deficiencies could be corrected by the expression of oncogenes like BCR-ABL, MYC, and NRAS (Henry et al. 2010). Aging leads to reductions in key parameters of cell fitness, promoting selection for oncogenic mutations that correct these parameters. In contrast, young progenitors exhibit the right amount of these same parameters (the evolved state). This state of being well-adapted means that oncogenes that boost these activities will actually be maladaptive—too much of good thing can be a bad thing, as seen in movement away from the peak in a fitness landscape. Excessive oncogenic activity can lead to clonal elimination, such as via apoptosis or differentiation. According to the current paradigm for oncogene-driven clonal expansions, formulated by Peter Nowell in 1976, each step in somatic evolution is limited by the occurrence of the oncogenic mutation, which confers a fitness advantage to promote clonal expansion. However, when oncogenes are expressed in normal healthy young stem and progenitor cell pools, the general result is a reduction in cellular fitness, leading to clonal exhaustion instead of expansion (DeGregori 2012). Clonal expansions require the right context to promote oncogenic adaptation.

Nonetheless, the occurrence of oncogenic mutations is still required to initiate clonal expansions, as described by Nowell, and increases in mutation burden in our somatic cells as we age will consequently contribute to cancer risk.

If a genetic mutation were adaptive simply by fixing cellular parameters, we would expect that such a mutation would restore cells' youthfulness instead of leading to cancer. Indeed, some mutations can be adaptive without being transforming (Bagby and Meyers 2007). However, the reason that oncogenes contribute to further cancer progression is that they confer additional phenotypes on cells that promote the hallmarks of cancer. Importantly, these additional phenotypes, such as genomic instability or invasiveness, need not be the ones that were adaptive.

Oncogenic selection driven by aged hematopoiesis represents adaptation in its simplest form: old age leads to reduced levels of key factors, and oncogene expression leads to restoration of youthful levels of these same factors. One could also envision how an oncogene could be adaptive not by restoring some factor, but by circumventing its effects. For example, above I described how p53 mutation provides resistance to hypoxia—p53 mutation does not restore oxygen levels but simply makes the cells more tolerant of low oxygen (Hammond and Giaccia 2005). The key point is that mutational adaptation, whether by correcting or circumventing a fitness-impairing trait, is promoted by microenvironmental changes. Just as the adaptiveness of the lactase gene mutation in human populations was entirely dependent on dairy farming and consumption of dairy products, the adaptiveness of oncogenic mutations is highly dependent on tissue context. Aging is a tissue context that cannot be avoided.

Identifying the Causes of Tissue Decline

To review, aging reduces the fitness of hematopoietic progenitor cells, including specific parameters that underlie a cell's ability to survive and divide, which leads to enhanced selection for adaptive oncogenes. Cellular fitness declines due to insufficient pressure for natural selection to maintain tissue fitness beyond the years of likely reproduction. Now let us consider these cellular fitness changes within an individual—the proximate factors underlying cell and tissue decline in old age.

First, we know that cells can accumulate damage to their DNA and other cellular structures over a lifetime, simply due to the wear and tear of living (White and Vijg 2016). Every time DNA is replicated in a cell, a small number of errors can be made, and these can be passed on to subsequent generations of cells. These errors can be made either in DNA, changing letters or even rearranging larger parts of the DNA, or in the epigenome. As described above, our DNA and the histone proteins attached to it are epigenetically modified, and these alterations greatly affect the expression of genes. Like replication of the DNA itself, the replication of these epigenetic modifications is imperfect, and thus errors are passed on to the next cell generation. Individuals with inherited syndromes resulting from impaired DNA repair exhibit premature aging, or progeria, although the extent to which accumulating mutations contribute to normal aging is still unknown. In fact, since most cell divisions happen early in life, while we do not exhibit substantial physiological decline until late in life (Figure 7.1), there are clearly mechanisms to limit the impact of accumulating mutations, such as by purging stem cells with phenotype-altering mutations. Finally, other structures in our cells, such as the pores in the membrane surrounding the nucleus in each cell, can accumulate damage over life, reducing the functionality of the cell. Reduced nuclear pore functionality particularly impairs long-lived and differentiated cells that no longer divide (D'Angelo et al. 2009). All of this cell intrinsic damage can reduce the evolved function of the cells and the tissues that they constitute. Some of these changes, at least the genetic and epigenetic ones, can also provide the mutations, including epigenetic changes, upon which somatic selection can act. A subset of these mutations could contribute to the oncogenic state.

Notably, the impact of these changes on somatic evolution is not necessarily through changes in the cells that will eventually form a cancer, as declining cells and tissues constitute the microenvironment for the cell that might have oncogenic ability. Changes to the cellular niche may be more important for age-dependent oncogenic adaptation than changes to the actual cell with oncogenic potential. In addition to changes within the cells of the tissue, older ages are associated with alterations in the structure and chemical makeup of the extracellular matrix and milieu, the latter of which includes growth factors and cytokines. Growth factors and cytokines relay signals between cells, with substantial impacts on the behavior and fate of the receiving cells. Even if a cell were to accumulate zero intrinsic damage over the individual's lifetime, it would find itself in a tissue that had undergone substantial age-related changes. These tissue changes will make the cell less adapted to its environment. Thus, age-related decline in tissues will promote selection for new cellular phenotypes adaptive to the new microenvironment. To draw an analogy with organismal evolution, we can think of these changes as akin to environmental perturbations throughout Earth's history. Our planet's changing landscape has been responsible for the drastic changes over time in the species that inhabit it. Similarly, our bodies' physiological decline in old age creates new opportunities for somatic evolution.

Aging-Associated Inflammation Alters Adaptive Landscapes

As we further consider physiological changes that can account for tissue fitness decline, and thus enhanced oncogenesis in old age, we should take into account factors that could lead to generalized and relatively synchronized decline throughout the body. One strong candidate is inflammation. Inflammation is a critical program that responds to pathogen invasion and tissue damage through a complex orchestration of various leukocytes and cytokines. Inflammation is known to increase in old age and is associated with a number of diseases, including heart disease, type 2 diabetes, lung diseases like chronic obstructive pulmonary disease, and—most important for our discussion—cancer (Goto 2008; Lisanti et al. 2011). Dampening inflammation with drugs like aspirin reduces the risk of heart disease and some cancers (but it also increases bleeding, ulcers, and other side effects) (Rostom, Dube, and Lewin 2007). Inflammatory programs have been under very strong selection, and genes required for inflammation are highly conserved across almost all animals. This strong conservation reflects the essential role of inflammation as the first line of defense against infections and in orchestrating tissue repair following damage. This critical program appears to go into overdrive late in life, contributing to disease by disrupting tissue and systemic homeostasis (Kotas and Medzhitov 2015). Since natural selection favors youth over old age, the benefits of this program for survival and thus reproduction in youth far outweighs its negative impacts in old age.

Inflammation has been long associated with cancer occurrence, but the impact of inflammation has primarily been considered as a promoter of cancer hallmarks, increasing proliferation, survival, and invasion of cancer cells (Todoric, Antonucci, and Karin 2016). In contrast, Henry and collaborators showed that inflammation is a key mediator of the aging-dependent functional impairment of B-cell progenitors in the bone marrow of mice (Henry et al. 2015). As discussed above, this fitness impairment with age leads to selection for oncogenic events that correct aging-dependent alterations in metabolism, gene expression, and signaling. Notably, inhibiting inflammation in mice prevents these aging-associated deficiencies in signaling and gene expression. Hematopoietic progenitors in old mice expressing anti-inflammatory mediators exhibited phenotypes that were closer to those of young progenitors—in other words, blocking inflammation essentially prevented functional aging of these cells. Strikingly, inhibiting inflammation completely prevents selection for the activated NRAS oncogene (Figure 8.2, bottom panel). Thus, by preventing aging-associated fitness reductions in B-cell progenitors, diminishing inflammation abrogates selection for adaptive oncogenic events. Without room for improvement, oncogenes are no longer positively selected. These results emphasize that while we may not be able to avoid oncogenic mutations, which occur as cells divide throughout our lives, we can manipulate microenvironmental factors to influence whether or not these mutations lead to cancer. While aging is very multifactorial, researchers may be able to identify individual factors that are key to oncogenesis.

Indeed, this study reveals that a specific alteration in the aged microenvironment—increased inflammation—may be a major driver of aging-associated leukemias as a result of altering selective pressures for oncogenically initiated cells (Henry et al. 2015). It is important to note that the study specifically compares oncogenesis in mice of different ages, from young to old. Despite the fact that cancer is largely a disease of old age, almost all cancer modeling in mice employs only young mice. Based on the mutation-centric view, the contribution of aging to cancer reflects the time for mutation accumulation, so providing or inducing the oncogenic mutations is believed to suffice to explain cancer. Contrasting with this view, the studies by Henry and colleagues show that the inflammatory aged tissue microenvironment is critically important in oncogenesis: the age and state of the host really does matter.

This study raises an interesting question: can modulation of inflammation limit aging-associated oncogenesis in humans? Indeed, regular aspirin intake has been shown to reduce the incidence of a number of cancers, although the side effects of daily aspirin administration complicate this potential prevention strategy (Rothwell et al. 2011). Finally, the implications of links between inflammation and cancer could be broader. For instance, studies by Louis Vermeulen, Douglas Winton, and colleagues showed that selection for mutational inactivation of p53 in colon stem cells is potentiated by colitis, an inflammatory disease of the colon (Vermeulen et al. 2013). Moreover, obesity is also associated with substantially increased cancer risk, and the increased inflammation associated with obesity could contribute to this association (De Pergola and Silvestris 2013).

Microenvironmental Perturbations in a Growing Tumor

As noted above, a major force sculpting evolution on earth has been life itself (Ward and Kirschvink 2015). For example, the evolution of photosynthetic organisms drove an increase in oxygen levels in the atmosphere. Oxygen was toxic to the organisms present before the advent of photosynthesis, and thus reduced the fitness of organism exposed to air. The increase in oxygen also resulted in selection for new species adaptive to this context, including more complex eukaryotic cells with mitochondria. Thus, life influences the evolution of other life.

In the same way, a growing cancer creates an ever-changing environment that spurs additional somatic evolution. The growing tumor creates microenvironmental challenges that must be overcome for further progression (Gatenby and Gillies 2008). For example, as the tumor grows, it quickly reaches a size whereby diffusion from neighboring blood vessels can no longer supply the tumor with enough oxygen. This size barrier engenders a potent selection for cells that mutationally acquire a more glycolytic phenotype. Our cells normally use oxygen for cellular respiration through mitochondria, as a way to convert sugars like glucose into energy. However, with insufficient oxygen, tumor cells can evolve to use glucose to generate energy, which reduces their need for oxygen. This glycolytic process also better produces cellular building blocks to support cell divisions (Vander Heiden, Cantley, and Thompson 2009). While the glycolytic phenotype may not have been advantageous for the same cell in a healthy tissue, this phenotype can become advantageous in a growing tumor. A number of oncogenic drivers, including BCR-ABL, RAS, and MYC, enhance glycolysis.

Interestingly, this acquisition of the glycolytic phenotype itself sets up the next barrier, as glycolytic cells produce a lot of lactic acid, which acidifies the microenvironment. This in turn selects for cells that are adaptive in this acidic environment, such as through mutations that prevent apoptosis in this context (Gatenby and Gillies 2008). Of course, we can also surmise that these barriers are never breached for many tumors, perhaps due to low numbers of cells within the tumor that provide for insufficient genetic variability upon which these selective hurdles can act. Indeed, as highlighted in Chapter 5, we know that many cells with oncogenic mutations and even early malignancies are common in our tissues, far outpacing the frequency of the advanced cancers associated with these same tissues. Thus, fortunately, the barriers to cancer development work—at least most of the time.

Somatic evolution requires the phenotypic diversity created by genetic and epigenetic mutations and is driven by alterations of fitness landscapes as we age. The pattern of late-life cancer incidence is dictated by the waning of evolutionary investments in tissue maintenance. Through adaptive oncogenesis, we can understand how evolutionary strategies at the species level have affected cancer susceptibility in our bodies. Alterations of fitness landscapes are not limited to aging but are evident within growing tumors, influencing selection for additional oncogenic mutations. Understanding cancer genesis and progression from the evolutionary perspective should facilitate the design of interventions to prevent and control cancer.

CHAPTER 9

Changing Adaptive Landscapes with Carcinogenic Exposures

WHILE AGING IS A MAJOR FACTOR in cancer development, roughly one-half to two-thirds of cancers owe their origins to external factors, including diet, exposure to air pollution, alcohol consumption, exposure to sunlight, gamma irradiation, chemotherapy, infections, and smoking (Stein and Colditz 2004; Anand et al. 2008). Just as the longer lives of modern humans has augmented the incidence of cancers of old age, modern exposures and lifestyles can lead to increased cancer risk. Humans have subverted the evolutionary pressures that shaped our physiology by modern living and by lifestyle choices. Evolution has not had nearly enough time to shape responses to high-intensity radiation, air pollution, or smoking, which have only affected our survival for the last handful of generations. According to the current paradigm, carcinogen exposures can increase the frequency of mutations, providing fuel for somatic evolution. This chapter will explore how extrinsic carcinogenic insults affect stem cells and their microenvironments. I will use the second principle of adaptive oncogenesis to argue that by promoting selection for oncogenic events that are not adaptive in undamaged tissues, changes in fitness landscapes are critically important for carcinogen-mediated oncogenesis.

Extrinsic Causes of Altered Fitness Landscapes

The Earth has not just been subject to "Earth-intrinsic" changes over the past 4.5 billion years. Outside influences, such as the strength of radiation from the Sun (which has changed over these billions of years) and impacts from objects in space, can clearly influence evolutionary trajectories. The most famous example occurred about sixty-six million years ago off the Yucatan Peninsula, when the impact of a giant meteor is thought to have been the final blow to the age of dinosaurs. The impact completely changed the environment of the entire planet, destroying by fire much of the Earth's forests, instituting a sort of nuclear winter, and destroying most of the plant life on Earth (Ward and Kirschvink 2015). These changes led, over time, to the evolution of new species adaptive to the new environments that reconstituted the Earth after the impact. Indeed, today's mammals can give thanks to this meteor impact, as it ushered in the Cenozoic era (also called the age of mammals).

We can of course draw an analogy to our own bodies. While the somatic evolution that leads to cancer late in life can in part be ascribed to the aging of our tissues, external exposures can substantially influence cancer incidence (Stein and Colditz 2004; Anand et al. 2008). The best-known examples are exposures to sunlight (ultraviolet light) and pollution and, of course, smoking. If you have ever seen pictures of a blackened and unhealthy smoker's lung, you can appreciate how this tissue would present a very different microenvironment for its constituent cells. The cells of our lungs evolved to function in the healthy lungs of a young person, at least through ages where reproduction was likely. When someone smokes, he or she severely disrupts that evolved relationship, creating a microenvironment that can foster the evolution of new phenotypes adaptive to this charred landscape (analogous to what happened after the meteor landed near the Yucatan Peninsula). Some of the cells that adapt to this new landscape could carry oncogenic potential, starting the cell clone down the path toward cancer. Cells that are adapted to the normal healthy lung are likely to be well-behaved and functional, but cells that adapt to the damaged smoker's lung are more likely to be bad actors. Adaptation to such an environment should select for phenotypes that diverge from evolved ones, including for resistance to damage-mediated cell death and other hallmarks of cancer. A nasty environment does not reward good behavior.

Smoking: Sabotaging Evolved Tumor Suppression

No other factor comes close to cigarette smoking in its impact on cancer deaths: smoking causes about one-third of all cancers (American Cancer Society, n.d.; Proctor 2012). Lung cancer is the leading cause of cancer mortality, with around 160,000 deaths attributed to lung cancers per year in the United States. There are about 225,000 cases of lung cancer per year, leading to a grim conclusion: most people diagnosed with lung cancer will die from their disease. Cigarette smoking causes about 85 percent of lung cancers. If one includes the many other associated cancers, lung diseases, and heart disease caused by smoking, tobacco accounts for about one in five deaths in the United States, or about 480,000 deaths per year.

Smoking is thought to cause lung cancer by increasing the frequency of mutations. Indeed, smoking leaves its fingerprint in the DNA of lung cancer cells. Carcinogens in smoke directly induce particular types of DNA mutations, and these specific changes in the DNA code can characterize mutations found in known oncogenes and tumor suppressor genes in smoking-associated lung cancers (Hainaut and Pfeifer 2001; Pfeifer et al. 2002). However, mutations can account for only a fraction of the large increase in cancer risk due to smoking. A recent report examined mutation burden in thirteen cancers associated with smoking, comparing mutation burden with similar cancers in people who had never smoked. Only lung adenocarcinoma and larynx cancer showed a substantial smoking-dependent increase in burden (4.5- and 2.5-fold, respectively). Two of the cancers showed less than 1.5-fold increases, and the other nine exhibited no increase in mutational burden in the cancers from smokers. Overall, the increase in mutations associated with smoking for all cancers combined was only 1.15-fold, or 15 percent (see Figure 9.1). Nonetheless, the authors concluded "that smoking increases cancer risk by increasing the somatic mutation load" (Alexandrov et al. 2016, 618).

Interpreting these data without the bias imposed by the prevailing paradigm results in a different conclusion: smoking substantially increases

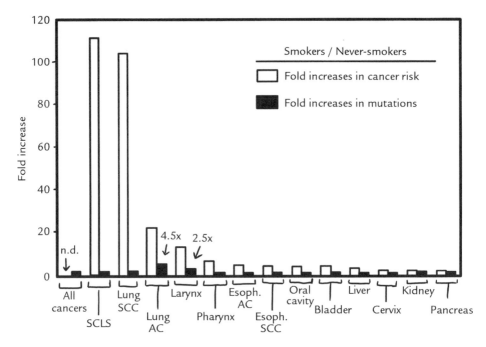

Figure 9.1. Smoking increases cancer risk without consistent increases in mutations. The increase in risk for the indicated cancers in smokers (including former smokers) compared with "never-smokers" is graphed with white bars. The increase in risk for all cancers combined was not determined (n.d.). The increase in mutations comparing the same cancers in smokers and never-smokers is graphed with black bars. The fold increase in mutations for small cell lung cancer (SCLC) is based on only three patient cancers from never-smokers, and firm conclusions for this cancer will require a larger sample size. SCC is squamous cell cancer. AC is adenocarcinoma. Esoph. is esophageal. Data were graphed from Alexandrov, L., et al., "Mutational Signatures Associated with Tobacco Smoking in Human Cancer," 2016. © Michael DeGregori

the risk of many cancers without substantially increasing mutation burden. These data are inconsistent with the current paradigm that links smoking to cancer almost entirely through increasing mutation rates. While carcinogens in cigarette smoke can increase cancer risk by acting as mutagens, smoking is clearly affecting other aspects of somatic evolution (Rubin 2002). Just as aging is associated with clear changes to tissue architecture and function, smoking is well known to be associated with dramatic changes to lung tissue, including inflammation and tissue damage, which cause great impairments in lung function that can manifest themselves as emphysema or chronic obstructive pulmonary disease (COPD). In fact, even among smokers, COPD and poor lung function are independent risk factors for the development of lung cancer (Sekine et al. 2014; Takiguchi et al. 2014). I argue that the massive array of changes to tissue structure and function brought about by smoking is underappreciated as a contributor to the development of cancers, and that the role of these factors in lung cancer development is largely unexplored.

Niles Eldredge (1999) describes how catastrophic events throughout the history of the Earth have led to reproducible patterns of evolution. Similarly, there is a pattern to smoking-driven somatic evolution in the lung. The fact that the same oncogenic mutations in lung cancers associated with smoking have been observed in thousands of individuals is indicative of this pattern: similar alterations in the microenvironment select for similar adaptive strategies. Mutations activating the KRAS oncogene are found in about one-quarter of smoking-induced lung cancers. These mutations are known to occur early in cancer evolution, and they likely reflect selection for KRAS activation in the damaged lung tissue. In addition, the fact that no two cancers possess exactly the same sets of mutations demonstrates that there are different adaptive solutions for similar problems within lungs caused by smoking, as well as that chance plays a role. Thus, smoking creates a fitness landscape with certain common features, but also with some features unique to each individual due to his or her genetics, environmental exposures, lifestyle choices, and chance.

Interestingly, lung cancers that develop in people who never smoked differ in the oncogenic events that drive their development, relative to the lung cancers of current and former smokers. For example, lung adenocarcinomas in nonsmokers frequently exhibit mutations in the EGFR gene that lead to deregulated signaling, while lung adenocarcinomas that develop in current and former smokers tend to have mutations in the KRAS gene. Furthermore, EGFR mutations are rare in lung cancers of smokers, and KRAS mutations are rare in lung cancers of nonsmokers. As is the case for many other cancers, the primary factor associated with lung cancers in nonsmokers is old age (Couraud et al. 2012). While it is possible that smoke carcinogens are more likely to induce KRAS mutations, evidence is lacking. The theory of adaptive oncogenesis supports an alternative explanation: changes in fitness landscapes in the lung are different for aging and smoking, which leads to selection for distinct sets of oncogenic events that are differentially adaptive to these contexts. Similar to how adaptation of a species to a change in climate would be expected to involve selection acting on different genes than adaptation to a change in pathogenic threats, differences in the lungs of smokers and the elderly could lead to selection for different adaptive mutational events.

While the key differences in fitness landscapes are unknown, smoking leads to massive accumulation of smoke particles, damage from constituent chemicals, and abnormal immune cell infiltration. Cigarette smoke induces both a potent inflammatory response in the lung and cycles of injury and repair (O'Callaghan et al. 2010). There is a direct correlation between markers of lung inflammation and airflow obstruction, and COPD is highly associated with inflammation. As use of nonsteroidal anti-inflammatory drugs (NSAIDs) like aspirin is associated with lower lung cancer risk, it is likely (but not yet shown to be the case) that inflammation is contributing to lung cancer initiation and/or progression in smokers (Rothwell et al. 2011). KRAS mutations in lung cancer lead to hyperactive signaling in the cancer cells, including through pathways important for the replication of DNA and the production of other cellular molecules necessary for cell duplication (Kim et al. 2005). Perhaps these same pathways are impaired in stem and progenitor cells in the lungs of a smoker, so that KRAS activation is adaptive. Just as inflammation impairs hematopoietic progenitor cell function in old mice, leading to selection for progenitors expressing RAS oncogenes (Henry et al. 2015), we can hypothesize that smoking-induced lung inflammation may also alter the tissue landscape to promote selection for RAS oncogenic activation (in this case, for the KRAS gene family member). Indeed, smoking has been shown to enhance tumor development initiated by KRAS mutations, at least in part by increasing lung inflammation (Takahashi et al. 2010).

While the lungs of the elderly also exhibit impairment and increased inflammation, the nature of these changes diverge from those in the lungs of a smoker, promoting selection for EGFR activation. We can theorize that fitness landscapes will differ in aged and smoked lungs, with different heights and positions of peaks, and with alternative paths up cancer peaks. The activation of KRAS and EGFR, respectively, may simply be more adaptive in the smoked and aged lung microenvironments.

This view of lung cancer genesis has important implications for possible prevention strategies. While there is currently no way to reverse oncogenic mutations that have occurred or prevent those that will occur with continued smoking, medical researchers could develop strategies to manipulate fitness landscapes. Screens for compounds or treatments that restore evolved lung phenotypes could yield new approaches to prevention and treatment. It may be particularly useful to reverse inflammation, which is associated with the lungs of both smokers and elderly nonsmokers (Sekine et al. 2014; Sharma, Hanania, and Shim 2009). Methods that lead to more rapid restoration of the youthful lung phenotype after a smoker quits, or that better maintain lung landscapes as we age, could reduce the ability of oncogenic mutations to be adaptive. Reduced adaptation would mean less expansion of the oncogene-bearing clone, with consequent reductions in the odds of further cancer development.

Irradiation Exposure and Programmed Mediocrity

Although responsible for far fewer cancers than smoking, the association of exposure to ionizing radiation and increased cancer incidence has been recognized for over a hundred years, although the underlying mechanisms remain poorly understood. X-rays and gamma rays are common types of ionizing radiation. Animal models clearly demonstrate that exposure to ionizing radiation potently induces cancers, mostly leukemias and lymphomas (Fleenor, Higa, et al. 2015; Kominami and Niwa 2006; Kaplan and Brown 1951). The most famous (and tragic) demonstration for the power of radiation exposure to induce cancers came after atomic bombs decimated the Japanese cities of Hiroshima and Nagasaki at the end of World War II, resulting in substantial increases in the incidence of multiple cancers, particularly leukemias. Radiotherapy is used to treat about 60 percent of solid tumors in the United States and is highly associated with secondary malignancies (Curtis et al. 1997). Since ionizing radiation is a potent mutagen, and cancer development requires the acquisition of mutations in oncogenes and tumor suppressors, the prevailing paradigm attributes cancer causation by DNA-damaging carcinogens like ionizing radiation to the direct induction of oncogenic events by these agents (Little 2000). However, species evolution is driven by environmental change, not by changes in mutation rates, and somatic evolution can be expected to follow similar rules.

Notably, radiation exposure—whether in children treated for malignancies or in atom bomb survivors—also results in long-term reductions in tissue fitness, such as in blood cell production (Fleenor, Higa, et al. 2015; Mody et al. 2008). Andriy Marusyk, Courtney Fleenor, and collaborators showed that prior irradiation exposure leads to programmed and reversible changes in HSC fitness, including reduced ability to support hematopoiesis resulting from stem cells' impaired ability to renew themselves (Fleenor, Rozhok, et al. 2015; Marusyk et al. 2009). HSCs from irradiated mice exhibited precocious differentiation down blood lineages, failing to maintain themselves as stem cells. Oncogenic mutations that impaired differentiation were selected for in HSC pools from previously irradiated mice—but not in those from unirradiated mice, due to the ability of these oncogenes to restore HSC self-renewal. So analogous to the oncogenic adaptation observed in aged progenitor cell pools, the poor self-renewal engendered by prior radiation exposure promotes selection for oncogenes that restore this self-renewal. Consequently, leukemias were induced more readily in the previously irradiated backgrounds by these oncogenes.

These studies suggest a model that can be used to explain tissue maintenance and tumor suppression following a DNA-damaging insult, depending on the extent of the damage. In natural contexts, such exposures would typically involve only a small fraction of stem cells. Stem cells receiving high doses of radiation exposure frequently undergo cell death via apoptosis (Shao, Luo, and Zhou 2014). This makes sense from an evolutionary perspective, as cells with a lot of DNA damage will likely be less functional and are also at greater risk for possessing oncogenic mutations. However, it is also important for stem cells to not overreact, which could lead to stem cell depletion. In the context of more modest damage, as modeled in mice above, stem cells receiving but surviving the insult activate a program that reduces self-renewal and increases differentiation

(Fleenor, Rozhok, et al. 2015). This programmed mediocrity, at least in the more natural context of damage to the occasional cell, facilitates the elimination of the damaged cell from the stem cell pool without creating a void, thus leading to a more gradual replacement of damaged stem cells with less damaged (that is, more fit) peers. In line with the theory of adaptive oncogenesis, maintaining the fitness of the stem cell pool is expected to suppress tumors by limiting selection for adaptive oncogenic mutations.

However, in the more modern context of radiation exposure (in clinical or accidental contexts, or following an atomic explosion), virtually all surviving stem cells may have experienced genotoxic stress, and thus the same programmed mediocrity will reduce the fitness of the entire stem cell pool, which can promote selection for adaptive oncogenic mutations (Fleenor, Rozhok, et al. 2015; Marusyk et al. 2009). Similarly, apoptotic mechanisms eliminating many stem cells after irradiation lead to selection for resistance to this apoptosis, and this resistance can be mediated by loss of the p53 tumor suppressor (Bondar and Medzhitov 2010; Marusyk et al. 2010). As far back as the 1930s, Alexander Haddow (1938) proposed that by inhibiting cell proliferation, carcinogens can promote the outgrowth of cancer-initiating cell clones that can resist this inhibition. Appreciating how radiation and other carcinogens alter mutational selection through negatively affecting stem cell fitness suggests a novel way to limit the risk of cancer following irradiation or other genotoxic insult: the restoration of stem cell fitness.

Infections and Cancer

Almost all organisms are in evolutionary battles with viruses. While viruses can cause diseases as diverse as the common cold and AIDS, they have also been shown to contribute to cancers. At least 15 percent of cancers are thought to have a viral cause (Ewald and Swain Ewald 2013). Viruses carry their own DNA or RNA genomes but depend on host cells for replication. To achieve this goal, some viruses push tissue cells into the cell cycle, to increase production of the enzymes and building blocks critical for manufacturing more viruses. These enzymes and building blocks would normally be used by the cell to duplicate itself but are instead commandeered by the virus. For this purpose, some viruses encode genes that stimulate entry and progression through the cell cycle (Levine 2009). Animals have evolved mechanisms to prevent and limit virus infections, such as triggers that induce cell death upon recognition that the cell is infected. Of course, viruses have evolved countermeasures, and they possess anti-apoptotic protein-encoding genes that facilitate the survival of the infected cell for long enough to allow for virus replication. In many cases, the infected cell will be killed in the process of producing virus. In other cases, the virus can persist in the infected cells, either in a latent form (not producing virus) or with low-level virus production that does not kill the enslaved host cell (Westrich, Warren, and Pyeon 2017). Still, a small fraction of virus infections can promote malignant progression, and in that case the pro-cell cycle and anti-apoptotic viral genes evolved by the virus to optimize viral production in cells can contribute to oncogenic transformation of these cells. The number of people infected with the viruses greatly outpaces the number that will develop the associated cancer, similar to the greater prevalence in our tissues of oncogenic mutations than of the cancers associated with these mutations. Even though viral genes can function as oncogenes, other hurdles thankfully stand in the way of cancer progression.

Paul Ewald and Holly Swain Ewald (2013) proposed that by delivering multiple oncogenic hits at once, viruses can allow cells to overcome multiple barriers to carcinogenesis. The authors argued that viruses are major causes of cancer, beyond the currently accepted 10–20 percent. The provision of multiple viral oncogenes could allow a cell clone to more easily cross a fitness valley in an adaptive landscape (Figure 6.1). Normally, cancer evolution proceeds through selection from randomly generated mutations, most of which will reduce fitness. Only a minor fraction is potentially adaptive. In contrast, viruses directly deliver oncogenes, which were evolved to maximize viral reproductive success even if not to cause cancer. The vast majority of infections serve the virus's evolved purpose— to reproduce and spread to other hosts. Cancers are not necessary to achieve these goals and are likely the result of mistakes during virus replication that benefit neither the virus nor the host.

By integrating the Ewald barrier theory with adaptive oncogenesis, we can envision how viruses and other pathogens can both deliver oncogenes and change microenvironments, with the latter favoring selection for viral and cellular oncogenic adaptations. Viruses can induce chronic inflammation and other changes to tissues that could alter carcinogenic risk. In other cases, the pathogen or condition that induces inflammation is distinct from the virus that provides the oncogenes. As a case study, let us first consider human papillomavirus (HPV) and cervical cancer. Certain strains of HPV are well-established causes of multiple cancers, including almost all cervical cancers and cancers of the anus, penis, and vagina (Fernandes et al. 2015). HPV is sexually transmitted. While throat and neck cancers have primarily been associated with tobacco use, with reductions in smoking an increasing fraction of these cancers are now attributed to HPV (transmitted during oral sex). The carcinogenic strains of HPV encode for potent oncogenic proteins, including E6 and E7 (Westrich, Warren, and Pyeon 2017). Among other functions, E6 and E7 lead to the inactivation of the host p53 and RB proteins, respectively—two of the most important tumor suppressor proteins in mammalian cells. The roles of E6 and E7 in the cancer-causing potential of HPV have been clearly established.

Cervical cancer is the most common of the HPV-associated cancers, and the second most common cancer in women worldwide. Far more women are infected with carcinogenic HPV strains than will develop cervical cancers, which indicates that other factors must be in play (Westrich, Warren, and Pyeon 2017). First, most of those infected eliminate the virus within a couple of years. However, in some women the infection becomes persistent, and for them reactivation of the virus can increase the risk of progression to carcinoma. Causing cancer does not appear to be part of HPV's evolved strategy. In fact, while HPV does not normally insert its genome into ours, the HPV-caused cancers possess pieces of the HPV genome integrated into the human host genome of cervical cells. Notably, these integrated pieces of HPV invariably contain the E7 and E6 genes and have lost much of the genome important for virus production. This event does not appear to serve either the virus or its human host. In fact, the integrated HPV frequently loses the viral E2 gene that would normally keep the expression of E6 and E7 in check.

Even the presence of persistent HPV infection does not guarantee carcinoma development, and in fact most affected women will not develop cervical cancers. Notably, infection with other sexually transmitted pathogens, like the herpes virus or chlamydia or gonorrhea bacteria, are associated with increased risk of HPV-induced cervical cancer. These secondary infections cause persistent inflammation and ulceration of the cervical epithelium and thus would drastically alter the microenvironment for somatic evolution (Fernandes et al. 2015). In fact, chronic inflammation is a well-established risk factor for cervical cancer. Moreover, smoking greatly increases the risk of HPV-associated cervical cancers, and smoking is known to induce systemic inflammation (Bozinovski et al. 2016). Ironically, a key mediator of adaptive immunity, B-cells, appear to contribute to the inflammatory response and consequently carcinogenesis initiated by HPV (Fernandes et al. 2015). Perhaps the chronic state of persistent HPV infection converts these normally antipathogen or anticancer lymphocytes into promoters of cancer progression.

As we have discussed, inflammation could promote carcinogenesis through a variety of mechanisms, including by promoting the survival of cancer cells or their more aggressive phenotypes. From the perspective of the theory of adaptive oncogenesis, I would argue that inflammation also alters the cervical tissue environment in a way that no longer favors the evolved type for epithelial stem and progenitor cells. This change could engender adaptation by HPV oncogenes or by the additional oncogenic hits required for the progression of cervical carcinoma. By providing the early oncogenic events, HPV greatly increases the odds of cancer development. However, alterations of the fitness landscape, such as those mediated by inflammation, are likely key determinants of whether these HPV oncogenes are adaptive. Such adaptation promotes the expansion of the oncogene-bearing cell clone, with consequent increased risk of further cancer progression.

While we cannot avoid the mutations that accumulate during the development, maintenance, and aging of our bodies, the knowledge that viruses can deliver oncogenic events suggests that limiting HPV infections should greatly decrease cervical cancer incidence. Indeed, the demonstration that HPV causes cervical and other cancers spurred the development of a vaccine to these dangerous strains of HPV, which has led to dramatic reductions in HPV infection in young women and girls and expected future decreases in HPV-associated cancers (Garland et al. 2016).

Another example of pathogen-associated cancer is the Epstein-Barr virus (EBV), which infects about 90 percent of humans at some point and can cause diseases like mononucleosis. While causing relatively minor or no problems in most people, EBV has been associated with Hodgkin and Burkitt's B-cell lineage lymphomas and a number of other lymphomas and cancers (Grywalska and Rolinski 2015). Notably, all of these cancers are relatively rare. As the EBV genome encodes several oncogenes, and almost all humans become infected with it, why do the vast majority of us never develop the associated cancers? Additional factors clearly contribute to cancer risk. For example, immune suppression appears to facilitate EBV-associated lymphoma development, which is more common in people with AIDS or those who have had immunosuppression associated with organ transplantation.

Notably, Burkitt's lymphoma prevalence is particularly high in equatorial Africa and Papua New Guinea, coinciding geographically with the incidence of malaria (Grywalska and Rolinski 2015). Infection with the malarial parasite *Plasmodium falciparum* could promote Burkitt's lymphoma development by stimulating the expansion of the B-cell population, which would provide more opportunities for subsequent oncogenic events, or by suppressing T-cell immunity. Further evidence points to a role for inflammation—whether induced by malaria, aging, or other causes—in lymphoma development. Inflammation could create a tissue microenvironment that might switch an EBV oncogene from being maladaptive to being adaptive. Finally, additional oncogenic events are clearly needed for lymphoma development, often involving amplification of the expression of the MYC oncogene. EBV is necessary, but not sufficient, for lymphoma development. EBV does not appear to provide everything needed for a cell clone to fully cross a fitness valley, but by introducing multiple oncogenes it can greatly increase the odds that this transit will be made if the right context (for example, inflammation or immune suppression) is present and additional events (such as MYC gene amplification) occur.

While Burkitt's lymphoma is rare in developed countries, chronic liver inflammation caused by long-term infection with hepatitis C virus (HCV) is a major contributor to hepatocellular carcinoma, a class of liver cancers. HCV promotes a chronic inflammatory microenvironment in the liver, which contributes to oncogenic progression (Levrero 2006). While these viruses also appear to encode genes with oncogenic activities, HCV-mediated damage to the liver involving inflammation, fibrosis, and cirrhosis appears to substantially contribute to cancer development (Lin, King, and Chung 2015). Inflammation associated with HCV infection can create liver dysfunction for decades, which can lead to cancer in a fraction of individuals. Virus-mediated destruction causes inflammation, and the high cell turnover needed to maintain the liver further increases the odds that oncogenic mutations will occur. Liver cirrhosis is a risk factor for carcinoma development, and contexts that exacerbate liver damage like other infections and excessive consumption of alcohol further increase risk. The more altered the tissue environment, the greater the promotion of cancer progression, likely involving oncogenic adaptation to this new environment.

For a nonviral example, colonization of the stomach with the bacteria *Helicobacter pylori* has been shown to greatly increase the odds of stomach cancers. Either eliminating the infection with antibiotics or blocking inflammation with NSAIDs can reduce this risk (C. Wu et al. 2010). Again, barriers are in place, as many more people have *H. pylori* in the stomachs than develop the associated ulcers, and most of these people with ulcers will not go on to develop stomach cancer.

In all, we can see how viral and other infections can contribute to cancer incidence by affecting the key evolutionary parameters of selection and mutation and by disabling key tumor suppressive mechanisms of their hosts. Some viruses directly provide oncogenes, thus partially obviating the need for mutations in host oncogenic drivers. Nevertheless, virus infection is clearly not sufficient for cancer formation, and contexts like chronic inflammation (which enables the development of cancer phenotypes and promotes oncogenic adaptation) or immune suppression (which overcomes the immune barrier) are required. Other pathogens (like *H. pylori*) appear to primarily promote cancer by acting as modulators of tissue microenvironment: for example, they may induce chronic inflammation, which in turn can alter selective pressures for oncogenic mutations. In this case, the carcinogenic context is provided, but not the oncogenic drivers.

Adaptive Oncogenesis: Unanswered Questions

A new theory should be judged by how well it explains currently available data and known mechanisms. In addition, it should continue to be challenged by further experimentation and observations. For a process as complicated as cancer development, a new theory is not likely to explain everything, nor will it supplant all previous explanations. In this light, let us consider some of the deficiencies and limitations of the theory of adaptive oncogenesis. First, we have discussed the many mechanisms that animals have evolved to limit cancer incidence through periods of likely reproductive success. These mechanisms include intrinsic checkpoints like apoptosis, hierarchical tissue organization, immunity, and tissue peer pressure, as well the stabilizing selection for the evolved phenotype that is a key principle of adaptive oncogenesis. It is clear that the tumor suppressive stabilizing selection, which eliminates cells with potentially oncogenic mutations, acts in concert with and in part through these other mechanisms. They are not mutually exclusive. Evolution has acted to place multiple hurdles in front, and throughout the course, of cancer development.

Similarly, the relative contribution of alterations induced by microenvironmental change to the fitness impact of mutations—the other critical principle of adaptive oncogenesis, underlying the substantial increases in cancer risk that accompany old age or following exposures from radiation, smoking, viruses, or other factors—has only been partially explored. Aging and these other insults affect mutation accumulation, immune function, and other parameters that clearly also contribute to oncogenic progression. Moreover, factors such as smoking and irradiation create a lot of cell death. The resulting homeostatic cell replication to restore cell numbers increases cancer risk, including by increasing the numbers of mutations that are inevitable with each cell division. Decimation of stem cell populations also increases the role of drift in these now small populations, which could lead to the fixation of oncogenic mutations that would normally be purged because they have some fitness cost. While I believe that there is significant experimental evidence suggesting that adaptive oncogenesis plays an important role in determining the pattern of cancer incidence (such as incidence late in life for aging-associated cancers), I by no means desire to discount the importance of additional changes associated with aging and other cancer causes. Our bodies are extremely complicated, as are the cancers that develop within them, so we should not expect simple answers.

Another limitation of the theory of adaptive oncogenesis is that we do not know the extent to which it applies to somatic evolution in solid tissues, from which the majority of cancers originate. Most of the experimental and computational work that my lab has performed has been in the hematopoietic system. While the theory is based on more general considerations and should logically extend to other tissues, I recognize this limitation. In particular, stem cell pools are organized very differently in the hematopoietic system than in solid tissues (as discussed in Chapter 10), which will affect the relative roles of selection and drift in somatic evolutionary dynamics. Testing the theory in solid tissues, while more difficult for technical reasons, is a major next step for my research group.

There are several major unknowns that prevent a full development of the theory of adaptive oncogenesis, and even a complete understanding of why we and other animals age. The evolutionary reasons for aging are clear, but we do not know nearly enough about the proximate reasons. How much does the accumulation of epigenetic and other mutations contribute to physiological decline, relative to the decline in other nongenetic cellular and extracellular components? Is damage accumulation in stem cells the most important factor, or does the decline of the differentiated and nondividing cells of tissues (including muscle cells, neurons, and cells that make up stem cell niches) play a more important role in aging and thus adaptive oncogenesis? How is somatic tissue maintenance modulated by natural selection to achieve advantageous life strategies for different species? Importantly, what molecular and cellular pathways are the most critical, how are they modulated, and can we capitalize on this knowledge for human benefit?

We know that increasing inflammation is a contributor to late-life physiological decline and increased cancers, but we do not understand why inflammation increases late in life. Does a positive feedback loop between tissue damage (due to waning maintenance) and inflammatory responses to this damage amplify inflammation late in life? Answers to these questions will be critical for fully formulating any theory of aging or cancer. Finally, we currently lack a quantitative understanding of the parameters of somatic evolution, such as the extent to which different contexts affect the fitness of particular stem cell populations and how these contexts influence the distribution of the fitness effects for mutations. We need to increase our understanding of somatic evolution to approach the quantitative understanding that we have obtained for population genetics and evolutionary biology. There is much to be done.

CHAPTER 10

Tissue Architecture and Tumor Suppression

THE PATTERN OF INCREASED CANCER risk late in life is remarkably similar for malignancies initiating in different tissues with very different numbers and organizations of stem and progenitor cells. I have discussed how this pattern implies the existence of a causative factor that is extrinsic to the oncogenically mutated cells and systemic in the body. One such factor could be the inflammation that accompanies our overall physiological decline in our postreproductive years. Nonetheless, differences in stem cell pools for different tissues should affect the relative roles of different evolutionary forces. As I have discussed, in large populations the roles of mutation and selection will dominate. The frequency of a genetic allele, such as one caused by a new mutation, in a large population will mostly be dictated by the impact it has on fitness. In contrast, in small populations, the role of drift in determining the fate of mutations can become important. The role of drift will be inversely proportional to population size and the magnitude of the fitness effect of the mutation. The differential numbers and organization of stem cells in tissues in our bodies should thus play a role in the character of somatic evolution and carcinogenesis.

Selection: Strength in Numbers

The human blood system originates from a single pool of 11,000–300,000 stem cells residing in the bone marrow (Abkowitz et al. 2002; Wang, Doedens, and Dick 1997). This number is debated, as it depends on how these hematopoietic stem cells (HSCs) are assayed, but 11,000 is likely the low-end estimate. As discussed above, this is a relatively small number given the hundreds of billions of cells produced by the hematopoietic system each day. From the perspective of accumulating mutations, this small number of stem cells creates a small target for mutation accumulation, particularly given the very low rate of cell division (about once a year) that these cells undergo. This pool of stem cells appears to be largely intermixing, in that HSCs enter the circulation (probably after dividing) and can take up residence in a new stem cell niche somewhere else in the body. A stem cell that resided in the bone of one big toe could end up in the other. The ability of stem cells to home to bone marrow niches forms the basis for the success of bone marrow transplantations, in which bone marrow from one individual can be intravenously delivered to reconstitute the hematopoietic system of another. Additional studies using parabiotic mice (two mice sewn together to create two animals with chimeric blood systems) showed that HSCs from one mouse ended up in the other (D. Wright et al. 2001). Thus, HSCs in a body can be considered as a single population.

Each time an HSC divides, the two daughter cells likely enter the circulation, where they compete with other divided stem cells for limited niche space in the bone marrow. We know that niche space is limited, since at least for most of an adult's life (whether mouse or human) the number of HSCs stays very constant, and these cells are highly dependent on their niche to maintain their identity (Morrison and Scadden 2014). A stem cell without its stem cell niche soon differentiates into a cell committed to follow the path of differentiation into a mature and typically short-lived cell. Maintaining precise control over cell numbers, whether in the blood system or any tissue, is a key component of tumor suppression. The evolution of multicellularity demanded order and conformity.

We can think of the dynamics of HSCs entering circulation in search of niche space as a biological game of musical chairs—except the number of chairs remains constant, each participant duplicates itself upon standing up, and most of the participants stay seated during any given round. This dynamic creates a competitive environment in which the fitness of each HSC is assessed in terms of its ability to survive in circulation, find a niche, and successfully occupy it. There is certainly also a role for chance (drift) in this process, although with enough HSC competitions, the most fit genotypes should predominate. With numbers of 10,000–300,000, selection will play a much greater role than drift, at least for mutations that have a larger effect on fitness than the inverse of the population size.

With this understanding for how HSCs compete for niche space, we can appreciate that cellular fitness is not just a measure of the ability of a cell to divide and survive. The cell must also be able to find and keep a niche spot. Numerous studies have shown that mutations that increase the entry of HSCs into the cell cycle actually reduce the ability of these cells to remain stem cells: despite increasing cycling, the net effect on cellular fitness will be negative, as the HSCs will be more likely to commit to differentiating (DeGregori 2012). Given that many oncogenic mutations increase entry into the cell cycle, this system is highly tumor suppressive. The best strategy for an HSC is to not divide too often, as this puts the cell at risk of losing its seat in the niche, but to divide occasionally to maintain clonal representation in the pool. Fortunately, this is also the evolved strategy. Evolution has created a beautiful system in which the best interests of the stem cell (maximizing its cellular fitness) are perfectly concordant with the best interests of the organism (maximizing its fitness by effectively producing blood cells and limiting oncogenic transformation).

According to the theory of adaptive oncogenesis, larger stem cell pools should strengthen stabilizing selection in youth, including the ability to eliminate HSCs acquiring oncogenic mutations that increase cell cycling. However, larger HSC pool sizes increase the power of positive selection to promote the expansion of oncogene-bearing clones following age-dependent reductions in HSC pool fitness. An HSC's fitness is dependent on its interactions with its microenvironment. Alterations of this niche in old age reduce the fitness of the HSC pool and lead to selection for adaptive mutations, some of which are oncogenic.

There are other tissues that are also maintained by large intermixing cell pools, and in them similar dynamics in terms of competition and selection will apply. In some cases, most cells of a tissue maintain the potential to replicate, and thus the bulk of the cells can be called upon to reconstitute the tissue when needed. The liver is organized in this way. One can remove two-thirds (two of the three lobes) of the liver from a rat, and the remaining lobe will grow to reform a liver of almost exactly the same size as the original, but now containing just one large lobe (Taub 2004). This is an amazing example of tissue homeostasis. While all cells in the hematopoietic system originate with the stem cells, reconstitution of the liver entails replication of all of the primary liver epithelial cells (hepatocytes). The synchrony of this cell replication is amazing, even at the biochemical level. Following the loss of liver, all of the hepatocytes enter the cell cycle, even simultaneously turning on the required enzymes for replication.

Given the critical importance of the liver for chemical detoxification, glycogen storage, and the production of serum proteins, this system clearly evolved to ensure the uninterrupted function of the organ. In this case, the regenerating cells are not stem cells in the classical sense but are the bulk of cells maintaining the tissue, and they number in the billions. Notably, there is evidence of designated stem or progenitor cells in the liver that function to generate mature hepatocytes in fetal development and perhaps under more chronic damaging contexts like viral hepatitis or alcoholic liver disease, but the number of these stem or progenitor cells should still be very large (Miyajima, Tanaka, and Itoh 2014). Selection should dominate over drift in these contexts, even with some restriction in cell movement. Similarly, the differentiated and functional exocrine cells of the pancreas, which produce and secrete digestive enzymes into the small intestine, appear to be reprogrammed into stem-like cells upon injury, to mediate regeneration of the tissue (Ziv, Glaser, and Dor 2013). For both liver and pancreas, the maintenance of such large pools of cells with substantial replicative potential has implications for oncogenesis, since these cells may serve as targets for mutation and selection. Selection, stabilizing in youth and positive in old age, should dictate the timing and character of somatic evolution.

Drifting in Small Pools

Our intestines perform a very useful set of functions—the digestion of food, absorption of nutrients into our blood system, uptake of water, and elimination of waste—all while protecting us from external invaders. Intestines are organized in a modular fashion, being constituted by millions of nearly identical smaller units (about ten million in the human colon). Each unit in the small intestine has a villus that extends into the intestinal lumen. These finger-like projections greatly increase the surface area of the small intestine, amplifying its ability to absorb water and nutrients. Each villus has cells for absorbing water and electrolytes, blood vessels for transporting absorbed nutrients, specialized cells for secreting protective mucus into the lumen, and other cells that make antimicrobial peptides. The epithelial cells of each villus are maintained by a small pool of 10–20 stem cells at the bottom of an invagination that dips down from the base of the villus, called the crypt (Figure 10.1A). The large intestine similarly organizes stem cells into crypts but has no villi. Given that competition appears to be largely limited to between stem cells within each crypt, such small numbers will result in a much larger role for drift in determining the persistence or expansion of cells of different genotypes in the crypt. Studies in genetically engineered mice have shown that for each crypt, just one of the original 10–20 stem cells will end up dominating the pool within a few weeks—by chance alone (Snippert et al. 2010; Lopez-Garcia et al. 2010). For these studies, scientists used mice with engineered genomes, with the random expression of one of multiple different color-producing genes in crypt stem cells. Using these "rainbow" mice, the scientists could show that each crypt would be occupied by a single color of stem cells within a few weeks. Since these colors have no impact on fitness (they represent neutral mutations), the researchers could conclude that drift was an important force in these stem cell pools.

Mutations that have large impacts on fitness, increasing or decreasing fitness by more than about 10 percent, will be influenced by selection. Still, we can appreciate that genetic drift will play a larger role in the somatic evolution and eventual cancer development in the intestines relative to another tissue like the hematopoietic system. The exact im-

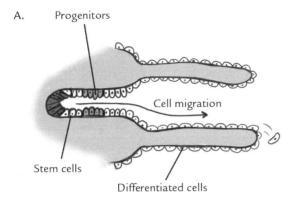

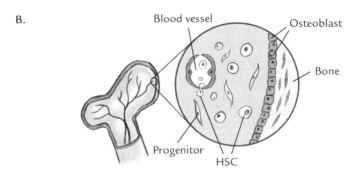

Figure 10.1. Stem cell pool organization in different tissues. A. Stem cells in the small intestine reside in the crypts below villi in groups of 10–20 cells. Stem cell divisions maintain the stem cell pool and also produce differentiated cells (via short-lived progenitors), which migrate up the villus. These cells are then shed into the lumen of the intestine. Drift dominates clonal dynamics in these small stem cell pools. B. Hematopoietic stem cells (HSCs) reside in the bone marrow in niches that include endothelial (blood vessel) cells, osteoblasts, and other cells. The HSC pool functions as one large population, as HSCs can migrate to other locations throughout the body (by entering and exiting blood vessels as shown). Selection dominates clonal dynamics in the large HSC pool. © Michael DeGregori

pact of this difference on oncogenesis is not known, but we can speculate that different oncogenic mutations will exhibit different probabilities of fixation in the two classes of stem cell systems. As described in Chapter 3, somatic mismatch repair (MMR) gene mutations are found in colon cancers but are rare in leukemias. Recall that MMR is an important system to repair mismatched nucleotides in DNA. Since these MMR mutations increase the frequency of new mutations by up to a hundredfold, they can contribute to tumor evolution. However, we also know that higher mutation rates come with a cost in terms of cellular fitness, as most mutations decrease cellular fitness. Let us assume that this cost is less than 5 percent. In that case, MMR mutations could persist in stem cells in the colon and become fixed even by chance alone within a crypt-bound pool, while the same mutation would be effectively eliminated in the much more competitive HSC pool.

The fixation of a MMR gene mutation in a colonic crypt pool would then increase the odds for subsequent mutations. At some point, one of these subsequent mutations might remove the restraint on stem cell pool size and overall intestinal architecture, and a small adenoma (a benign mass of cells) could form. The process of tumor evolution could continue, if the right mutations occurred in the mass to overcome the various hurdles that restrict cancer progression. Fortunately, in most cases these hurdles are not cleared, and the adenoma never develops into a carcinoma. However, in less lucky people, or those whose genetics or lifestyles (such as obesity or a diet high in fat) increase colon cancer risk, a carcinoma can develop, with additional evolutionary events leading to an invasive phenotype, with eventual dissemination into the blood stream and metastasis (Calle and Kaaks 2004).

Since the number one risk factor for colon cancer is advanced age, it is likely that changes in the colon with age contribute to the ability of oncogenic mutations to overcome hurdles (National Cancer Institute 2014; DePinho 2000). While an initial mutation might be fixed in a crypt by chance alone, it is also likely that selection plays a role. In line with the theory of adaptive oncogenesis, should the oncogenic mutation be adaptive in the old colon microenvironment (perhaps because of dietary and other factors), the odds of fixation increase, and more crypts will become occupied by oncogenic clones. Thus, the odds that the next oncogenic mutation will occur are accordingly amplified. Drift and selection are not at all mutually exclusive, and both forces are very relevant in the same tissue (such as the colon) for carcinogenesis. In fact, given that the age-related pattern of incidence for colon cancers is very similar to that for acute myeloid leukemia (National Cancer Institute 2014), which initiates in HSCs, age-dependent microenvironmental decline appears to similarly dictate oncogenic somatic evolution in these cancers with very different stem cell pool organizations.

The Evolution of Different Stem Cell Organizations

While a small stem cell pool increases the odds of conversion of the whole pool to a single genotype (clonality), it has an advantage in terms of cancer suppression. The expansion of the stem cell pool with an oncogenic mutation is limited by a pool size of 10–20, and thus the risk for the next mutation remains low even if cells with the oncogenic mutation dominates this small pool. In contrast, a mutation that increases the fitness of an HSC can lead to the expansion of a cell clone numbering in the thousands. We can only speculate why some systems evolved the large pool strategy (minimizing the role of drift, but facilitating cell expansions driven by adaptive mutations), while other systems evolved the strategy of having many small pools (increasing the role of drift, but restricting expansions). These differences could reflect the constraints that emanated from the ancestral systems (discussed below) or the different needs of these tissues: perhaps one system is better at making circulating cells, and the other at producing a static barrier.

In addition to the influence of stem cell organization on oncogenesis, evolved schedules for tissue turnover should be considered. Intestinal stem cells are busy, in that the epithelial cells making up most of the lining of the intestines are replaced every 3–5 days. Our intestines represent a critical barrier to the outside world and are bombarded with dangerous substances that we consume, such as toxins in our food, or that are generated within the intestines. These cells also interact with the trillions of bacteria in each of our large intestines as part of a delicate symbiosis—we protect and nourish them, and they provide us with essential vitamins, help us digest lipids, and facilitate our water absorption (Rooks and Garrett 2016). However, we also ingest many microbes with less friendly intentions, and the intestinal epithelial layer, together with many immune cells, forms a protective barrier. Turning over our intestinal epithelium on a regular basis may be important for keeping our intestines functioning well.

We can also appreciate how this system would be very useful from the perspective of tumor suppression. Cells of the intestines are on the front line, and many toxins are potentially carcinogenic, such as by causing damage to DNA. Fortunately, intestinal epithelial cells are essentially on a conveyor belt, produced by stem cells at the base of the crypt, differentiating into different functional cells, and performing these important functions as they make their way up the villus within the small intestines or out onto the epithelial lining of the colon (Figure 10.1A). Importantly, these cells are all destined for disposal within a week of their cellular birth, being shed into the intestinal lumen. Cells in our bodies that are on the front lines in terms of interacting with the outside world, such as the epithelial cells of our intestines and skin and the neutrophils of our immune system, tend to be short-lived. Neutrophils are white blood cells essential for eliminating bacteria and other invaders that make it past epithelial barriers (Kruger et al. 2015). Neutrophils suffer the cost of such warfare, including exposure to the toxic and DNA-damaging hydrogen peroxide that they themselves produce. These cells only live a few hours on average. Even if one of these various short-lived types of cells experiences an oncogenic mutation, such as due to the noxious exposures from the outside world, it is destined for disposal— whether shed from the skin, eliminated during defecation, or otherwise recycled in the body. A dead cell poses no risk.

The challenge of maintaining frontline cells, which experience the damages inherent in their protective roles but without facing the unacceptable risks of generating cancers, may be further accomplished by another factor that characterizes all of the associated stem cell populations: protective localizations (Goodell, Nguyen, and Shroyer 2015). The HSCs of vertebrates are sequestered in perhaps the most protected location in the body, inside our bones—close enough to blood vessels to facilitate nutrient delivery but far enough to limit damage from circulating oxygen (Figure 10.1B). While the lumen of the colon is an extremely inhospitable place (full of bacteria and wastes), colonic stem cells are hidden down in the crypts. Finally, even the stem cells of our most extensive barrier to the outside world, our skin, are localized in the basal (lower) layer of the skin, with some stem cells tucked under hair follicles (providing a bit of shade perhaps?).

The importance of these localizations for avoiding cancers in the respective tissues is at present purely hypothetical, as we cannot easily conduct the appropriate experiments. Even if we could somehow engineer mice whose colon stem cells were located on the surface epithelium, we would be changing so much about the stem cell niche that it would be impossible to know which factor was responsible for any potential change in cancer incidence. So in addition to the low number of stem cells (which presents a small target for mutation occurrence) relative to mature cells produced, the low rates of cell division (when most mutations would otherwise occur), the maintenance of a microenvironment favoring the evolved type (at least through youth), and the myriad other intrinsic and extrinsic tumor suppression mechanisms discussed above, vertebrates sequester their somatic stem cells for these tissues with high turnover in protected locations to minimize exposures to damaging agents. The selective pressures to limit carcinogenesis through reproductive years has been extremely strong throughout evolutionary history, and we can learn a lot about cancer by better understanding why we and other animals are so good at not getting it.

Tumorigenesis and Evolutionary Constraints

Different tissues evolved different means of maintaining and restoring themselves following injury, with large intermixing stem cell populations in some and many small isolated populations in others. These systems are clearly useful and functional, and their evolution likely relates to their adaptive value. Nonetheless, evolution does not create from scratch the same system that would result by design. Perhaps surface epithelial tissues evolved from primitive structures that used the small stem cell pool design, with structures like the human intestines representing a massive amplification of the modular crypt-villus unit. Natural selection works by modifying structures and functions already at hand, and the intermediate stages typically need to confer adaptive value. Thus, a major constraint on evolution is that it must create new adaptive innovations based upon what is already present, often in response to new environmental challenges and opportunities. Regardless, we can see that these tissues serve clear adaptive values now, and that they evolved not only to be useful but also to limit oncogenesis. Since we and other vertebrates rarely get cancer before old age, these different systems with different stem and progenitor cell dynamics are similarly able to limit somatic evolution.

As John Cairns and others have proposed, the need to avoid tumorigenesis has constrained animal evolution (Cairns 1975; Leroi, Koufopanou, and Burt 2003; Crespi and Summers 2005). The development and maintenance of every tissue and organ system must occur using a strategy that confers sufficiently low cancer risk, at least to the point that minimally diminishes reproductive success (Thomas et al. 2016). Animals did not need just to evolve an intestinal tract that was a good food-digesting and -absorbing organ with efficient protection from invaders, they also needed to evolve a gut whose architecture was consistent with low rates of carcinogenesis through reproductive years. There may have been other evolved solutions to the first set of problems (digestion, absorption, and protection) that were insufficiently robust for the second (tumor suppression), and thus these other systems were not evolutionarily successful strategies. The theory of adaptive oncogenesis proposes that any tissue must be designed so that stem and progenitor cells, by virtue of their interactions with other tissue components, are well adapted to their tissue niches. The need to maintain strong stabilizing selection within stem and progenitor cell pools through reproductive periods has constrained the evolution of tissue organization and maintenance strategies. Even though the strength of stabilizing selection will vary depending on the size of the susceptible cell pool, this mechanism should still function to limit oncogenesis through youth.

The recognition of cancer-driven constraints in our bodies has additional implications for academia and education: the specializations in the biological sciences may be detrimental. Have education and research become too differentiated? A developmental biologist cannot really understand the development of an organ without appreciating how the threat of cancer affected the evolution of its design and function. Moreover, a cancer biologist cannot appreciate cancer evolution without a good grasp of how tumor suppression requires the development of certain tissue architecture and maintenance strategies. These strategies promote the stabilizing selection that reduces the fitness value of oncogenic mutations. Perhaps most importantly, all biologists must possess a deep understanding of evolutionary mechanisms, including the adaptive value of the system that they study, the constraints that limited the system's evolution, and the roles of selection and drift in the underlying somatic genetics.

CHAPTER 11

Peto's Paradox

Different animals exhibit an amazing variance in life spans, from the roundworm *Caenorhabditis elegans* (which lives 2–3 weeks) to the ocean quahog, a mollusk that holds the record for animals (507 years) (Kenyon 2010; Butler et al. 2013). The vertebrate longevity record currently belongs to the Greenland shark: females reach sexual maturity at the spry age of 150 years, and the oldest individual found was about 400 years old and five meters long (Nielsen et al. 2016). Whatever their life spans, in all animals natural selection will have acted to greatly limit the development of function-damaging malignancies for as long as needed to maximize reproductive success. For mammals, body sizes and life spans also vary enormously, from a 20-gram house mouse with a maximum life span of 3–4 years to a 137,000-kilogram blue whale thought to live at least 100 years (Figure 11.1). Given that bigger animals have more cells (not bigger cells), we can estimate that the blue whale has almost seven million times more cells than a mouse. Not only must the whale produce more cells, but it must maintain and replace them for thirty times as long. This difference in the numbers of somatic cells can be further extended across the tree of life. For example, mice are very big and very long-lived compared to worm *C. elegans*, with each adult worm composed of only about a thousand somatic cells

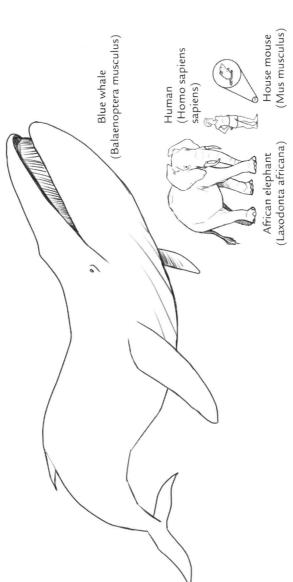

Blue whale
(Balaenoptera musculus)

Human
(Homo sapiens sapiens)

African elephant
(Laxodonta africana)

House mouse
(Mus musculus)

Figure 11.1. Peto's Paradox. Body size varies almost ten-million-fold and potential life spans vary more than thirtyfold among mammals. Yet all of the mammals shown similarly avoid cancer through years of likely reproduction. © Michael DeGregori

(Kaletta and Hengartner 2006). With such huge differences in cell numbers and life spans, how do all of these animals similarly avoid cancer through natural life spans in the wild, facilitating their reproductive success? While we know almost nothing about cancer incidence in Greenland sharks and blue whales, we can assume that natural selection has acted to limit cancer for most of their long lives. Richard Peto first contemplated this problem, and it therefore bears his name: Peto's Paradox (Peto et al. 1975). Two different classes of evolved strategies could allow for bigger bodies and longer lives: those unique to animals on particular branches on the tree of life, and those differentially employed across all animals.

Group-Specific Strategies for Tumor Suppression

While there are basic strategies that all animals use to avoid cancer, such as oncogene-induced apoptosis and immune surveillance (as described in Chapter 4), different subgroups of animals appear to have evolved their own solutions to Peto's Paradox to accommodate longer lives and/or bigger bodies.

Tumor suppressor genes are critical roadblocks to uncontrolled expansion and transformation of somatic cells. Increasing the number of copies of tumor suppressor genes has been suggested as an evolutionary strategy to prevent cancer in larger and longer-lived animals (Caulin and Maley 2011). For example, the elephant genome has twenty copies of the tumor suppressor gene p53, and its diploid genome has about forty p53 alleles, compared to the usual two for most mammals (Abegglen et al. 2015). Analyses of other related mammals demonstrated that this increase in p53 genes was limited to elephant species. These extra copies were formed by transposition, whereby extra gene copies can be inserted into other parts of a genome. The extra copies have mutations that lead to truncation of the protein, which lops off the normally critical DNA-binding domain. Additional studies demonstrated that the truncated p53 proteins are functional, enhancing apoptosis induced by DNA damage (Sulak et al. 2016). These results provide a potential way to explain how these extra p53 copies could boost tumor suppression through better elimination of cells with potential oncogenic mutations.

Nonetheless, Carlo Maley and colleagues analyzed numbers of tumor suppressor gene copies in thirty-six mammalian species, finding that increased numbers do not correlate with body size across these species (Caulin et al. 2015). Thus, multiplication of tumor suppressor genes is unlikely to be a general mechanism for suppressing cancer to allow for the evolution of longer lives and bigger bodies.

The naked mole rat has been endowed by natural selection with an exceptionally long life span among rodents (over thirty years) and has a very low cancer incidence (Tian et al. 2013). Given the relative lack of predators and other external hazards in the rats' sealed underground burrows, natural selection favored prolonged investment in tissue maintenance because it paid off in terms of reproductive success through these long lives. In terms of cancer prevention, naked mole rat fibroblasts secrete a special form of the molecule hyaluronan, a major component of the extracellular matrix in many tissues that controls cell proliferation and migration. The hyaluronan produced by naked mole rat cells is of larger molecular weight than that in humans and mice, and it accumulates at high levels throughout the rat's tissues. Cell culture experiments showed that manipulating hyaluronan could dictate transformability: removing this high molecular weight hyaluronan made the usually cancer-refractory naked mole rat cells susceptible to transformation to a more cancer-like phenotype (Tian et al. 2013). The type of hyaluronan determined how cells responded to their neighbors, with the larger hyaluronan form in naked mole rats leading to cell division arrest when cells contact other cells. If these mechanisms function similarly in tissues during oncogenesis, species-specific differences in hyaluronan could contribute to differential cancer susceptibility. Notably, this mechanism so far appears unique to naked mole rats.

As discussed above, cancer appears to be rare in most species in the wild. Most humans and of course laboratory mice, in which cancers and aging are largely studied, are no longer in the wild (Hochberg and Noble 2017). Only in modern or protected environments is survival typically limited by the maximum physiological life span of a species. As shown in the example in Figure 5.2, cancer incidence appears to be the mirror image of physiological decline (with survival under protected conditions used as a proxy). Thus, a simple explanation of the rare occurrence of cancer in wild animals and those brought into labs but whose life has not been extended beyond that achieved in the wild may be that they mostly do not survive to the ages where they would experience the trials of old age, such as diseases associated with tissue decline and cancer.

Understanding group-specific mechanisms for avoiding cancer can provide insight into the evolution of different life spans and body sizes within orders and families of animals, and it could also highlight strategies for clinical interventions. Nonetheless, given that all vertebrates and many other animals have tackled a similar problem (how to avoid cancer despite billions of cells and years of living), even if to varying degrees, and given the common physiologies of all vertebrates and certainly all mammals, there must also be shared strategies for cancer avoidance.

Tunable Tumor Suppression

When most cancer researchers think about the mechanisms through which some animals have evolved bigger bodies with longer lives, but without a proportional increase in cancer incidence, their attention turns to mutation rates (Caulin et al. 2015). Given the predominant paradigm holding that cancer incidence is limited by mutation occurrence during life, the focus on mutation rates is expected. Nevertheless, as discussed above, altering mutation rates does not appear to be a common strategy across the animal kingdom. The best data on mutation rates is for germline mutations, and larger animals—which tend to have smaller effective population sizes—have higher mutation rates than smaller animals (Lynch 2010). Indeed, the lowest mutation rates are found in single-celled organisms like bacteria and yeast, as their large population sizes allow for the efficient elimination of any mutation that even slightly compromises DNA maintenance and thus reduces fitness. However, information on somatic mutation rates in animals is much sparser and has mostly been derived from cell culture models. We do not know how cell culture, which involves much greater exposure to oxidative damage (including to DNA), influences mutation accumulation.

We also need to consider that the various DNA repair mechanisms will be differentially useful for different animal lifestyles. For example, some studies have shown that the efficiency of DNA excision repair, used to repair damage induced by exposure to ultraviolet (UV) light, is lower for rodents than for primates—at least for fibroblasts exposed in culture to UV light (Cortopassi and Wang 1996; Promislow 1994). However, UV-induced excision repair may be less relevant for small nocturnal mammals. With this caveat, these studies have claimed higher repair activity for UV-induced lesions in longer-lived and larger mammals. Could reducing mutation accumulation be a mechanism that has been tuned by natural selection within mammals to limit cell- damaging and potentially DNA-transforming mutations from accumulating during life? Since cells in culture may not properly reproduce processes that happen within animals, and fibroblasts are not the cells of origin for most cancers, we need much more data on mutation accumulation in human and other animal tissues. A few recent studies are starting to fill this gap.

One study demonstrated that individual stem cells from human liver and small and large intestine each accumulated roughly 2,500 mutations in a lifetime, while cancer incidence was much higher in the large intestine than the other two tissues (Blokzijl et al. 2016). Thus, the number of mutations accumulated correlates poorly with cancer incidence for different tissues within humans. While data for mice are similarly limited, a recent report did show that stem cells in the large and small intestines of mice accumulated roughly 250 and 500 mutations in a lifetime, respectively (Behjati et al. 2014). Thus, each human intestinal stem cell accumulates 5–10-fold more mutations relative to similar cells in the mouse, a difference that may be related to the roughly 25-fold longer potential life span for modern humans relative to lab mice. Considering that human intestines should have greater than 1,000-fold more cells, the cumulative mutation load for the human large intestine is roughly 10,000-fold greater than that for a mouse large intestine! From the perspective of Peto's Paradox, we can consider humans to be well-studied representatives of large, long-lived mammals, with mice representing small, short-lived mammals. This is obviously a small sampling of mammals. With this caveat in mind, the evolution of longer lives and bigger bodies apparently did not necessitate reduced mutation accumulation in tissue stem cells.

In most cancers, multiple mutational or epigenetic events are required to produce the cancer phenotype. Human cells may be harder than mouse cells to transform toward cancer. If so, and if this differential requirement for mutations dependent on size and life span extended to other mammals, it could represent an evolved mechanism to limit tumorigenesis. By requiring more steps, somatic cells from larger or longer-lived organisms like humans could be more resistant to progress toward cancer, as that progression would be less probable and would take longer. Indeed, Robert Weinberg and colleagues showed that freshly isolated mouse fibroblasts can be transformed into cancer-like cells in culture with only two oncogenic mutational hits, while human cells require up to six (Rangarajan et al. 2004). However, matters are not so simple. For example, it appears that the generation of retinoblastoma in mice requires more tumor suppressor gene disruptions than is the case in humans (Cobrink 2013). At present, there is insufficient evidence to either support or refute the idea that somatic cells from larger long-lived mammals require more oncogenic events to become cancerous than cells from smaller short-lived mammals.

As discussed above, telomeres protect the ends of eukaryotic cell chromosomes, and the loss of these protective ends leads to cellular senescence. The progressive shortening of telomeres, which can be countered by the telomerase enzyme, provides a cellular clock that can limit the number of cell divisions—an important cell intrinsic tumor suppressive mechanism. Telomeres also shorten due to oxidative stress, which could further function to weed out damaged cells. In human cells, telomerase activity is turned off in adult tissues (although the activity persists to some extent in stem cells), and thus cell replication leads to telomere diminution and eventual senescence.

Several groups of researchers have proposed that shorter telomeres and repression of telomerase have evolved in larger and longer-lived animals as a strategy to limit the increased cancer risk provided by the greater number of required cell divisions. Indeed, the repression of telomerase activity in adult tissues negatively correlates with body size in rodents (Seluanov et al. 2007). Larger rodents like capybara turn off or turn down telomerase activity in somatic cells in adulthood, while smaller rodents like mice maintain telomerase expression in somatic cells throughout life. Another study across a broader swath of more than sixty mammals also showed that shutting down of telomerase is associated with larger body size, and it also demonstrated that telomerase length is negatively correlated with longevity (Gomes et al. 2011). Thus, alterations in telomere maintenance may help solve Peto's Paradox, with greater restriction on excessive cellular expansion via telomere shortening in larger and longer-lived mammals.

Notably, the described associations between telomeres and longevity are not universal. Studies of *Glis glis*, the edible dormouse (so named because these rodents were considered a delicacy by ancient Romans), demonstrated that telomere length in oral epithelial cells actually increases at older ages (Hoelzl et al. 2016). These dormice are long-lived relative to many other rodents, surviving to about thirteen years, and telomere elongation may promote somatic maintenance during the older years when reproductive odds are still favorable. We can consequently surmise that telomere shortening with age is not required for tumor suppression in dormice. In addition, studies have shown that longer-lived birds and mammals exhibit slower telomere attrition during life (Haussmann et al. 2003; Dantzer and Fletcher 2015). These results are again consistent with the need to maintain somatic tissues, but inconsistent with the idea that reduced telomere maintenance as a tumor suppressive mechanism evolved to facilitate longer lives. Even within a species, such as Alpine swifts, the rate of telomere attrition within each bird is a predictor of life span, with slower attrition predicting longer life (Bize et al. 2009).

Data from humans is harder to interpret, given the difficulty of longitudinal studies and the impacts of lifestyle on telomere shortening (promoted by smoking, obesity, inflammation, and other stresses) (Muezzinler, Zaineddin, and Brenner 2013). Questions have been raised about whether telomere shortening is really a cause of aging (Simons 2015). Evidence about its links to cancer are also mixed, with mice engineered to lack telomerase experiencing increased cancer in some contexts and decreased cancer in others (Blasco 2005; Rudolph et al. 1999). Regardless, humans with inherited defects in telomere maintenance clearly exhibit premature aging and increased cancer susceptibility. Telomere attrition has been shown to promote chromosomal rearrangements, which could contribute to cancer risk. Through the lens of adaptive oncogenesis, we can further hypothesize that telomere shortening and the resulting genomic instability (due to the loss of protected chromosomal ends) in older ages could contribute to cancer risk, at least in part by altering adaptive landscapes via impairment of stem cells and their niches. In particular, telomere attrition can activate DNA-damage signaling and cell death or senescence, which should promote selection for reactivation of telomerase and for oncogenic mutations that prevent such cellular demise (Gunes and Rudolph 2013). Alternatively, telomere shortening with cellular divisions or stress could limit oncogenesis, by pushing oncogenically initiated cells into a genomic crisis that leads to cell death or senescence. Overall, it is difficult to derive a simple rule to link alterations in telomere maintenance strategies across species with the evolution of different life spans and body sizes.

I have discussed how hierarchical tissue organization, with large numbers of mature cells maintained by much smaller numbers of stem cells, is an important evolved tumor suppressive mechanism. Having fewer stem cells reduces the risk of an oncogenic hit occurring in a cell capable of initiating cancer. At least the number of hematopoietic stem cells (HSCs) does not scale with body size. Janis Abkowitz and colleagues (2002) estimated that the number of HSCs in mice and cats, and perhaps in humans, is similar (about 11,000 per individual) across these animals of very different body sizes. If body size can increase without a concomitant increase in stem cell number, then cancer risk may remain somewhat independent of body size. Still, the size of the human HSC pool is controversial, and we also need to consider that other stem cell pools will be unlikely to show similar constant sizes. For example, the skin and intestines are maintained by millions of segmentally organized epithelial stem cell groups, and it is hard to envision how evolution could increase surface area for these tissues without a proportional increase in these geographically distributed stem cell populations.

To figure out why cancer does not scale with body size and life span, we also need to consider anticancer mechanisms that could easily scale with increasing body size and longevity and / or could be modulated without excessive cost. For example, the immune system should naturally scale with body size, as a bigger animal has proportionally more immune cells capable of eliminating tumorigenic cells. Finally, the mechanism in the first principle of adaptive oncogenesis, stabilizing selection in young healthy tissues, also should not lessen with larger body size. In fact, stabilizing selection becomes stronger as population size increases, which minimizes the role of drift: the fittest persist. Thus, an oncogenically initiated cell in the large tissue of a whale should possess a lower capacity to overcome competition with the large number of nonaffected cells than would be the case in a mouse tissue. Such stabilizing selection is expected to counteract the increased risk of occurrence of oncogenic mutations due to the greater number of cell divisions required for larger bodies and longer lives. This mechanism can explain a well-described linkage: cancer incidence typically scales to natural life spans, with the vast majority of cases occurring beyond reproductive periods. The same can be said about aging phenotypes in general: they occur in postreproductive periods. Tissue maintenance would delay aging phenotypes in general and cancer in particular, at least according to the theory of adaptive oncogenesis. Simply having larger cell pools (including those for stem cells) and investing in the maintenance of these pools and their niches for longer should limit oncogenesis.

Unraveling a Paradox

Aging is not simply the passage of time, as it varies between species based on evolved maintenance strategies and among individuals within a species based on genetics, lifestyle, environment, and chance. Cancer follows the same rules. Considering extrinsic and integral mechanisms of tumor suppression provides a possible explanation of why longer-lived and larger animals do not suffer more cancer. We can then see how Peto's Paradox exists only in the context of the conventional theory holding that cancer is limited by the accumulation of mutations during life. Given that rates of aging and the late-life pattern of cancer appear to go hand in hand for different species, through the theory of adaptive oncogenesis I propose that there is an underlying common mechanism evolved by all animals to both delay aging and prevent cancer. According to the theory, doing the former will accomplish the latter. Tissue maintenance strategies can be considered as a knob that can be cranked up or down depending on what makes the best life strategy for reproductive success. Also according to the theory, common mechanisms would have a much more ancient origin, being conserved across metazoans. Importantly, since according to the theory, cancer suppression is a by-product of somatic maintenance, it is not necessary to postulate the existence of evolutionary innovations for this tumor suppressive mechanism.

The importance of raising paradoxes is that they stimulate new hypotheses and experiments to test them. Peto did this beautifully, and dozens of studies and publications from various labs have sought to address the paradox he proposed, as described above. While the theory of adaptive oncogenesis (specifically, stabilizing selection in stem cell pools during youth) may obviate the need for his paradox, time and much more experimentation will determine the extent to which group-specific mechanisms and the more general somatic tissue maintenance strategy explain how larger and longer-lived animals avoid cancer through their reproductive years just as small and short-lived animals do.

More boldly, a logical extension of the theory of adapative oncogenesis is that the evolution of animals necessitated the stabilizing selection mechanism proposed in the theory's first principle. The ability to eliminate cells that acquire phenotype-changing mutations is critical for the fitness of cells of a tissue as well as for eliminating cells with potentially oncogenic mutations. This mechanism was likely critical for all animals, from the smallest worm to the biggest whale, but it may have been particularly necessary for the evolution of larger and longer-lived animals. I have described how the evolution of multicellularity did not involve reductions in mutation rates to avoid function-disrupting or oncogenic mutations. Instead, strong stabilizing selection in stem and progenitor cell pools in animal tissues leveraged the same competitive mechanisms that can maintain the fitness of a population of single-celled organisms. Animals also evolved new mechanisms that attached costs to oncogenic phenotypes, such as apoptosis or the loss of self-renewal. The maintenance of high fitness for stem and progenitor cell pools, and thus good tissue function and tumor suppression, requires strategic investment in tissue maintenance to the point that maximizes reproductive success.

CHAPTER 12

Prolonging Tissue Maintenance to Delay Aging and Cancer

Humans have long sought the fountain of youth, and the fear of our own mortality has left its mark on our psyches, culture, religions, and art. As Oscar Wilde wrote in *The Picture of Dorian Gray*, "How sad it is! I shall grow old, and horrible, and dreadful. But this picture will remain always young. . . . If it were only the other way! If it were I who was to be always young, and the picture that was to grow old!" (1891, 34). While we cannot all have a portrait of us that ages instead of us, we can ask whether it is otherwise possible to delay the physiological decline associated with aging. Aging is not simply chronological, with some mammals showing age-related physiological decline after their first year and others only after more than a century. Just as tissue decline is mostly relegated to the latter part of potential life spans, so are most cancers: the evolution of longer lives goes hand in hand with delayed cancer incidence. I have discussed how the landscape of well-maintained (youthful) tissues limits oncogenic adaptation that can lead to cancer. If rates of physiological decline can be molded by natural selection, perhaps they can also be manipulated by humans. Accordingly, the extension of somatic maintenance strategies should not only postpone aging but should also reduce and/or delay the changes to tissue fitness landscapes that promote oncogenic adaptation and cancer.

Anti-Aging Strategies: Learning from Evolution

Humans are not exceptional, and we can learn a lot from studying evolved life strategies for other mammalian species. However, it is doubtful that we will find the secrets underlying extended life spans in some animals by comparing mice to whales, which have far too many genetic and phenotypic differences for us to tease out the ones that are important for longevity. A better strategy may be to compare changes in rates of senescence (of the individual, not just its cells) for subpopulations within a species, such as the opossums discussed in Chapter 1. However, even this comparison could take a large-scale effort, requiring the genome sequencing of many individuals from both the faster-aging mainland and slower-aging island groups of opossums, given that one would need to distinguish interindividual differences from interpopulation differences. Wild populations of animals exhibit substantial interindividual genetic variability just as humans do, which makes it difficult to identify the genetic alleles responsible for particular traits. Nonetheless, studies of wild animals do reveal that tissue maintenance strategies are modifiable through a relatively small number of genetic changes, given the rapidity of the evolution of altered life strategies. Studies of model organisms further substantiate the ability to manipulate longevity programs, which are typically inbred and thus genetically nearly identical. These lab studies have provided substantial insights into the genes and pathways that control rates of aging. While relevant variables are harder to control, studies in humans provide further support for the idea that dietary and pathway alterations can affect rates of aging.

Reversing aging seems like an insurmountable task, and indeed— given the substantial changes involved, from the extracellular matrices to cells and tissues—it may be nearly impossible. Moreover, it is doubtful that we can prevent aging. A better goal may be to improve tissue maintenance to slow aging, which would entail intervention before extensive physiological decline. Keeping us healthy longer, even without extending the maximal human life span, is not only a more realistic goal, but achieving it would actually reduce the burden on humankind. In contrast, having longer lives with more years spent in poor health would do the opposite.

According to adaptive oncogenesis, the best way to prevent or delay cancer may be to mitigate the physiological decline of our tissues, whether caused by aging or by environmental factors. Can we slow the age-dependent changes in fat accumulation, circulation, inflammation, repair capacity, tissue function, immunity, and so forth? Can we similarly alleviate changes associated with environmental exposures? How much could interventions that mitigate these changes in tissues also decrease the risks of the aging and exposure-associated diseases and organ failures that lead to our demise? While we can choose not to smoke, and we can typically avoid other carcinogenic insults including too much alcohol consumption, excessive sun exposure, and obesity, we cannot avoid aging. As Walt Kelly's Pogo said, "We have met the enemy, and he is us." Even if we do everything right in life, we are fighting another part of us: our evolution.

With the notable exception of modern times for humans and domesticated or laboratory animals, energy has been scarce and hard to come by. For example, a major cause of death for wild mice is cold, reflecting the inability of the mouse to burn sufficient calories to maintain its body temperature (Berry and Bronson 1992). Tissue maintenance is expensive, whether through replacement of worn-out organelles, DNA repair, antioxidant defenses, or eliminating and replacing damaged proteins. Natural selection has been able to alter these investments to maximize fitness in a particular environment. The investment in tissue maintenance wanes in older ages proportionally to reduced expectations of successfully contributing to reproduction. As demonstrated through the example of the opossum, and through the model organisms to be discussed below, the rapid evolution of life span diversity within closely related groups reveals that the dynamics of aging and tissue fitness decline are highly pliable via selection at the organismal level.

Like all animals, humans have evolved some programs that are important for our fitness in youth but that promote physiological aging. A process known as antagonistic pleiotropy is at work, whereby selection favors genotypes that are advantageous early in life, even if disadvantageous late in life, given the greater reproductive value of younger animals. Antagonistic pleiotropy refers to phenotypes with conflicting impacts on the organism. The highly conserved and intricate pathways required to mediate inflammatory responses are a great example of antagonistic pleiotropy, as inflammation is critical for eliminating infections and for wound healing, which clearly trumps their contributions to diseases like cancer and heart disease late in life (Kotas and Medzhitov 2015). We must keep such pleiotropy in mind when considering the design of antiaging and anticancer interventions: blocking an evolved pathway will not only temper its impact on the undesirable phenotypes (aging and cancer) but will also interfere with the traits associated with the selection for this program (for example, fighting infections). We do not wish to block aging pathways only to die of an infection.

Restricting Calories for Longer Life

In the 1500s, Alvise Cornaro (2014) of Venice proposed that reducing food intake would increase health and life span. He followed his own prescription, reportedly consuming 350 grams (about 12 ounces) of food and 414 milliliters (about 14 ounces) of wine per day and living about a hundred years. While historically interesting, the reporting of one person about his own life does not make for good science. Nonetheless, in the 1930s, Clive McCay and colleagues (1935) showed that reducing the food available to rats led to substantially increased life spans. Since then, research has shown that dietary restriction extends the life span across the tree of life to include yeast, flies, worms, rodents, and rhesus monkeys (Fontana, Partridge, and Longo 2010; Le Couteur et al. 2016). Life span can be extended by more than 50 percent in mice and worms. Caloric restriction entails reduced caloric intake to typically about 70 percent of what the intake would be without restriction (eating at will). Not only do the animals live longer, but they remain healthier by various measures of organ function and metabolism. Of note, one study showed that caloric restriction extended the life span of rhesus monkeys, while another did not. Regardless, both studies showed that "health span" (the fraction of life during which health is good) was extended. Calorically restricted monkeys exhibited better maintenance of organs (liver, heart, kidney, and so on), better brain and motor function, and more youthful overall metabolic profiles. The metabolic profile refers to various measures of proper metabolic health, such as levels of sugar, electrolytes, lipids, and cholesterol, and to how the body deals with an influx of sugar.

Limited fasting appears to have similar benefits in humans, lowering the incidence of metabolic syndrome, cardiovascular disease, and cancers at older ages (Fontana, Partridge, and Longo 2010; Le Couteur et al. 2016). Metabolic syndrome is characterized by a chronically poor metabolic profile, including poor control of blood sugar levels, as found in people with type 2 diabetes. As is well known, excessive food consumption, particularly when dietarily imbalanced, reduces longevity in people, and obesity is associated with increased cancer risk. However, being significantly underweight also has negative consequences on health and survival for humans, which is substantiated by underfeeding studies in experimental animals.

Food availability can be quite variable, and animals have evolved to respond to changes in this resource. The adaptation to environmental change is not the genetic adaptation that I have discussed, in which heritable changes in genes lead to the selection of phenotypes better adapted to a new environment, but adaptation within a single individual's life. Of course, this trait of adaptability, which can modulate somatic maintenance programs in response to resource availability, was itself favored by natural selection. This trait improves individuals' odds of surviving changing conditions, which are inevitable in the wild. Conditions of poor food availability promote a strategy of greater investment in tissue maintenance, with concomitant delayed reproduction, in anticipation of better times for reproduction in the future (but see Adler and Bonduriansky 2014). In essence, the individual is following a more "slow" life strategy, as discussed in Chapter 1, but within a single life span.

Experimental studies of caloric restriction have led to substantial advances in our understanding of the genes controlling longevity. Cell signaling mediated by insulin and insulin-like growth factors is the key pathway of responses to nutrient availability. It makes sense that modulating this pathway would allow an animal to alter its somatic investment strategy to match its environment, which is in good measure dictated by food availability. When blood sugar (glucose) levels are high, such as after a meal, special cells in the pancreas respond by making more insulin, which is secreted into the bloodstream. Insulin in turn binds to cells throughout the body to stimulate glucose uptake and metabolism. The failure to make sufficient insulin underlies type 1 diabetes, an autoimmune disorder that can lead to dangerously high blood glucose levels. A related protein, insulin-like growth factor (IGF1), is made by the liver and also orchestrates cellular responses to nutrient availability, including sugar and protein utilization and cell, tissue, and body growth (Figure 12.1). Thus, proper tissue adjustments to nutrient availability are critical components of healthy physiology.

Insulin or IGF1 signaling activates different pathways, including the mTOR pathway that is a key regulator of glucose metabolism and pro-

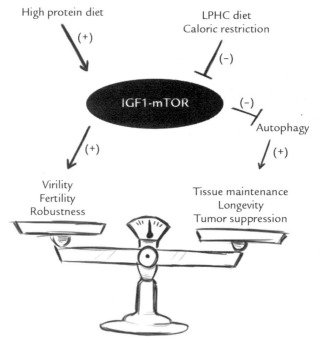

Figure 12.1. The IGF1-mTOR pathway and longevity. A high protein diet stimulates IGF1-mTOR signaling, leading to greater virility, fertility, and robustness. IGF1-mTOR also inhibits autophagy (discussed below) and reduces long-term tissue maintenance, tumor suppression, and longevity. Thus, a high protein diet increases fertility at the cost of longevity. The low protein–high carbohydrate (LPHC) diet and caloric restriction both inhibit the IGF1-mTOR pathway and therefore promote longevity and tumor suppression at the expense of fertility and virility in youth. © Michael DeGregori

tein synthesis. When food is restricted, such as with fasting, insulin levels in the blood and IGF1 or mTOR signaling in cells are reduced. Notably, in experimental models, genetic mutations that constitutively reduce IGF1 or mTOR signaling lead to many of the same life-extending benefits as caloric restriction. Yeast and animals (flies, worms, and mice) with such genetic mutations live longer and healthier lives, with reduced age-associated decline in overall physiology. Interestingly, single DNA base differences in the IGF1 gene appear to underlie a substantial amount of the size variation among dog breeds, leading to lower IGF1 expression in small breeds, and could also explain why smaller breeds live substantially longer (Sutter et al. 2007). There is even an example for humans (Fontana, Partridge, and Longo 2010). A mutation in the growth hormone receptor, which leads to reductions in the expression of the IGF1 receptor, in a small population of Ecuadorians leads to dwarfism, with lower risk of diabetes and cancer (but without increased longevity). Interestingly, certain genetic alleles that lead to lower IGF1 or IGF1 receptor levels are found in centenarians, suggesting (but not proving) that human life can be extended by dampening this pathway. Thus, lowering IGF1 signaling genetically appears to extend the life span across the eukaryotic tree of life.

Importantly, caloric restriction in mice lowers cancer risk (Fontana, Partridge, and Longo 2010; Le Couteur et al. 2016). Mice that were subjected to caloric restriction not only live about 50 percent longer but also exhibit 50 percent reductions in the rate of cancer, and the remaining cancers develop later in life. Studies of rhesus monkeys have reported a similar reduction (roughly 50 percent) in cancer incidence. Moreover, mice with growth hormone mutations that dampen IGF1 signaling similarly exhibit dwarfism and reduced and delayed cancer development, in line with observations of the Ecuadorians with familial dwarfism. Reduced cancer incidence is very consistent with the theory of adaptive oncogenesis theory: caloric restriction results in prolonged maintenance of more youthful tissue landscapes, which should impede oncogenic adaptation. While evolution across generations can change the life strategy for aging, which responds to long-term external challenges, caloric restriction of an individual activates tissue maintenance programs that similarly delay aging and cancer within its lifetime. In either case, prolonging tissue maintenance reduces the odds that oncogenic mutations will be adaptive, thus lowering cancer rates.

Given that altered gene alleles for IGF1 and other genes appear to lead to longer and healthier lives, with less cancer, why haven't these genotypes that reduce insulin signaling spread through all animal populations? The reason is that there is a cost associated with such genotypes, and this cost likely reduces reproductive success. For example, being smaller or less robust (that is, less muscular) could reduce competitive success with mates or survival from predators. In fact, the dwarf mice have shorter lives when housed with nondwarf littermates, which dominate their less robust dwarf siblings (Flurkey, Papaconstantinou, and Harrison 2002). The cost to reproduction and youthful size and vigor clearly outweighs the potential benefit of a longer life. In essence, the insulin pathway mutations, whether in flies, mice, or humans, results in the continuous execution of a program normally only instituted during caloric restriction, as opposed to the evolved ability of an individual to adapt appropriately to food availability.

Caloric restriction is also not without its costs (Fontana, Partridge, and Longo 2010; Le Couteur et al. 2016). In addition to the discomforts of hunger, caloric restriction can prevent wound repair, and weight loss can be excessive. Immune function can be impaired, increasing susceptibility to infections by bacteria, viruses, and worms. Caloric restriction can also decrease physical activity, since fewer calories are available to expend on exercise. In fact, some researchers question whether caloric restriction will actually result in life extension in the wild, given these costs (Adler and Bonduriansky 2014). Libido is also reduced, consistent with a strategy to delay reproduction until food is more plentiful. So individuals who voluntarily follow caloric restriction will go through life feeling hungry, with low energy and reduced sex drive, but they may live longer (if an infection doesn't kill them). Notably, there are drugs that can reduce insulin or mTOR signaling without the need to reduce food intake. One such drug is rapamycin, an inhibitor of mTOR. Numerous studies have shown that rapamycin treatment can extend the life span in multiple organisms (from yeast to mice), coinciding with reduced cancer incidence (Fontana, Partridge, and Longo 2010; Le Couteur et al. 2016).

Could drugs like rapamycin provide longevity benefits without the side effects? The answer so far appears to be no. Rapamycin is known to be immune suppressive, which may have limited ramifications for mice maintained under near-sterile conditions but could have real-life consequences. Rapamycin-treated mice also exhibit hallmarks of diabetes, testicular degeneration, and cataracts. However, there are indications that these undesirable effects of rapamycin are due to the inhibition of one mTOR protein (TORC2), while the beneficial longevity-promoting effects are via a related protein TORC1 (Lamming et al. 2012). Drugs that inhibit only TORC1 are being developed, and further experiments will be required to see whether it is possible to get the benefits of modulating insulin signaling without the costs. In any case, we will need to be careful before translating associations made in lab mice, maintained under clean and protected conditions, to people, who still need to optimally respond to infections and injury and to maintain good activity levels.

Finally, researchers have asked whether intermittent fasting, followed by full access to a normal diet, can confer some of the same health benefits as continuous caloric restriction (Brandhorst et al. 2015). Investigators devised a fasting-mimicking diet (FMD) for mice that involves four days of severely reduced caloric intake, with one day at about 50 percent and three days at about 10 percent of normal calories (the diet is designed to still provide sufficient essential nutrients like vitamins). Following the return of the mice to full and unrestricted access to mouse chow, the weight loss experienced during the FMD was quickly reversed, and over time there was no difference in cumulative food intake between the mice on the FMD and controls with continuous access to unlimited chow. Twice-monthly application of the FMD starting at what might be considered late middle age for a mouse (sixteen months) led to substantial improvement in multiple parameters of health at older ages, including maintained metabolic function, more youthful immune profiles, lowered visceral (midline) fat, and retarded bone density loss. The FMD group also was better at maintaining motor coordination, learning, and long-term memory into old age.

Importantly, cancer incidence in the FMD mice was reduced by 45 percent and delayed to older ages, coinciding with reduced inflammation. I would argue that the activation of a prolonged tissue maintenance program on the FMD not only delays physiological aging but also supports fitness landscapes that promote stabilizing selection. The evolved phenotype remains at a local fitness maximum, limiting selection for oncogenic mutations.

The mice on the FMD regimen also lived 11 percent longer on average, but the maximum life span was not extended, reminiscent of recent human gains in average but not maximum longevity. Notably, the same investigators performed a small study with human volunteers, using a three-month FMD that involved five days per month at about 40 percent of normal caloric intake (Brandhorst et al. 2015). Although these volunteers were healthy young adults, the FMD did positively affect several health parameters, resulting in lower blood glucose (a key metabolic parameter), lower markers of inflammation, and visceral fat loss without loss of bone density. Overall, these studies indicate that total caloric intake need not be reduced to increase health span and even lower cancer incidence. However, most people are unlikely to adopt a diet involving such regular fasting periods.

Dietary Protein and Our Health

Studies of caloric restriction have typically reduced all nutrients to the same extent. More recent studies have shown that a low protein–high carbohydrate (LPHC) diet maximizes longevity from flies and mice to humans, even when food intake is not restricted (Le Couteur et al. 2016; Mirzaei, Raynes, and Longo 2016). Mice were fed diets that varied in their ratio of protein to carbohydrates, but they were free to eat as much as they wanted. The diets that maximized longevity contained just under 10 percent protein. This LPHC diet also resulted in improved metabolism and cardiac function and more youthful immune profiles, but it led to weight gain and increased body fat. The mice ate more to obtain more protein from the provided low-protein chow. On the other hand, a diet with a 1:1 ratio of protein to carbohydrates maximized fertility, as assessed by sperm count and testes size in males and ovarian follicle number and uterine mass in females (Figure 12.1). Thus, a fivefold increase in protein content (from 10 percent to 50 percent) over the level that maximized longevity was required to maximize fertility.

In humans, analyses of multiple studies show that a LPHC diet is associated with positive health outcomes, lower cancer incidence, and longer lives, correlating with lower circulating IGF1 levels (which should reduce mTOR activation). However, the lower protein diet was actually associated with higher mortality for people over sixty-five, possible relating to poor protein absorption in the elderly. Negative impacts on the elderly, together with potential costs of the LPHC diet in terms of fertility, highlight the importance of considering individual circumstances in dietary strategies.

When animals are fed a low protein diet, they seek out high protein foods—animals from flies to humans have an amazing ability to modulate what they eat to attain a certain amount of protein in their diet. The balance sought reflects a much higher protein level than the one that maximizes longevity (Le Couteur et al. 2016). The higher protein diet maximizes fertility, so it is not surprising that evolution has led to cravings and behaviors to attain the protein intake that maximizes reproduction, even if at the cost of reduced potential longevity. Males will seek to attain even higher dietary protein levels than females, despite the long-term costs. For maximizing fitness, virility trumps longevity. These evolved programs make sense. Given the importance of sufficient protein for reproduction, when high protein food is not available, reproduction is delayed and a greater investment in somatic maintenance is made to improve reproductive odds later (when with luck a better diet will be available). So even if food is plentiful but has high carbohydrate and low protein content (like most fruits), reproduction will be delayed until a more appropriate food source is available.

The LPHC diet leads to reductions in mTOR activation (Figure 12.1), as also seen with caloric restriction, but without restricting calories. Amino acids are the building blocks of proteins, and a certain class of amino acids with branched chains can activate mTOR signaling. Reduced levels of branched-chain amino acids in the circulation of animals on LPHC diets likely dampens mTOR signaling, thus promoting tissue maintenance and longevity. It is important to note that branched-chain amino acids are more abundant in animal protein than in vegetable protein. Based on human studies, diets higher in vegetable protein than animal protein confer life span benefits, including heart health and lower rates of type 2 diabetes, providing yet another reason why you should eat your vegetables (Le Couteur et al. 2016).

The LPHC diet leads to higher body weight and greater fat in mice, which is surprising given that these mice exhibit better cholesterol profiles, healthier hearts, and longer lives. Being plump (up to a limit, of course) is not necessarily incompatible with good health. While exercise can be difficult and even harmful under caloric restriction, exercise may be able to reduce the body weight and fat gain associated with the LPHC diet. The impact of combining the LPHC diet with more exercise on life span, health, and cancer incidence has not been determined. It would make sense that excess calories consumed and increased body fat acquired on an LPHC diet would provide the energy required for more exercise, which would in turn reduce weight and fat gain. So maybe we can have our cake and eat it too, as long as we go for a jog afterward.

Cellular Recycling, Longevity, and Cancer

Caloric restriction and mutations or drugs that reduce insulin signaling all affect a common cellular process known as autophagy. Autophagy, from the Greek for "self-eating," is a process that involves the digestion of cellular constituents from within the cell. Autophagy eliminates cellular trash, mediating the disposal of protein aggregates and the turnover of organelles (like mitochondria) that become unable to function well (Green and Levine 2014). Inhibition of autophagy leads to the accumulation of dysfunctional organelles and other structures, which can greatly reduce cellular functionality. Autophagy also represents a recycling mechanism, as it breaks down degraded organelles and structures. The resulting simpler building blocks—sugars, amino acids, and lipids—are used again for cellular functions and maintenance.

Autophagy has been shown to be reduced as we age. As direct evidence that autophagy contributes to longevity, mutations in model organisms like flies and worms that reduce autophagy lead to shortened life spans, and genetic manipulations that enhance autophagy have been shown to increase health spans and life spans (Madeo et al. 2015). Median life span was extended by about 15 percent in genetically engineered mice with higher autophagy in their tissues, and these mice remained leaner and had better control of glucose levels (Pyo et al. 2013). Importantly, in flies and worms, the life-extending benefits of caloric restriction and insulin pathway mutations are completely dependent on autophagy, as mutationally impairing this mechanism abrogates these benefits (Figure 12.1). Conversely, mutations that reduce autophagy in mice lead to more cancer (Takamura et al. 2011). Thus, autophagy constitutes a major antiaging and anticancer process.

Autophagy appears to be a common strategy across the animal kingdom to maintain tissue landscapes (Green and Levine 2014). Could this be the mechanism that is tuned by natural selection to modulate the investment in tissue maintenance to maximize the likelihood of reproductive success? Does higher constitutive autophagy also contribute to delayed cancer incidence concordant with delayed senescence? Delaying aging is highly associated with delaying cancer, not only across the animal kingdom but also in laboratory models. We know that autophagy can directly promote the maintenance of stem cell populations. Moreover, the state of nondividing support cells is extremely important for the maintenance of cells with replicative potential, like stem cells (Scadden 2014). The accumulation of damaged lipids, proteins, and organelles in nondividing cells can decrease the ability of these cells to support stem cells, resulting in poor adaptation of stem cells to the age-damaged niche. As humans and other animals age, autophagy appears insufficient to mitigate the accumulation of damaged cellular constituents. However, enhancing autophagy prolongs the youthfulness of tissue structure and function, which should also sustain a fitness landscape that disfavors oncogenic change.

Finally, that there is always a cost for evolved life extension. The cost of enhanced autophagy in youth is likely through the energetic demands of organelle and large molecule turnover. In fact, the autophagy-enhanced mice described above were leaner despite similar consumption of food—while this may sound like a dream come true for many people, having to eat more to maintain similar body mass would be a questionable evolutionary strategy under conditions of rare survival to ages where the benefits of extended maintenance could be realized (Pyo et al. 2013). Still, for evolved delays in senescence onset, energy could be invested in enhanced autophagy and tissue maintenance (and thus tumor suppression), instead of reproduction, if external conditions favored this strategy.

Dampening Inflammation

To review, chronic inflammation is associated with aging and has been causally linked with a number of age-related diseases, including cancer and heart disease. Genetic variations found in humans that live exceptionally long lives (centenarians) are often in genes involved in inflammation, with effects on these genes consistent with reduced inflammation contributing to extended longevity (Salvioli et al. 2009). Regular treatment with nonsteroidal anti-inflammatory drugs (NSAIDs) like aspirin has been shown to substantially reduce the incidence of several cancers by as much as 40 percent (Rostom, Dube, and Lewin 2007). NSAIDs are particularly effective for preventing colon cancers, which are highly associated with inflammation. NSAIDS have also been shown to reduce clonal evolution in Barrett's esophagus, which is associated with higher risk of esophageal cancer (Kostadinov et al. 2013). Most recently, a large clinical trial of more than 10,000 individuals demonstrated that treatment with an antibody that sequesters the pro-inflammatory cytokine interleukin-1β substantially reduced cancer incidence, with a several fold reduction in lung cancers (Ridker et al. 2017). Moreover, I discussed direct evidence that experimentally reducing inflammation can prevent the selection for otherwise adaptive oncogenic events. By maintaining more youthful tissues, quelling inflammation can limit the aging-associated landscape changes that promote oncogenesis. Interestingly, caloric restriction and autophagy have both been shown to reduce inflammation (Hursting et al. 2013). However, at this point, we do not know the extent to which dampening inflammation contributes to the abilities of caloric restriction and autophagy to extend the life and health spans.

Of course, it's not so simple: we need inflammation. We must weigh the benefits of NSAIDs versus the costs, which include increased risk of stroke and gastrointestinal complications like ulcer bleeding. NSAIDs like aspirin can block platelet function required for clotting (Rostom, Dube, and Lewin 2007). Moreover, inflammation is an essential program for combating infections and repairing damaged tissue, and the responsible genes have thus been under strong positive and purifying selection. In fact, in the trial referred to in the previous paragraph, treatment with interleukin-1β sequestering antibodies increased the risk of death by infections (Ridker et al. 2017). Consequently, overall mortality was unaffected by antibody treatment for participants in this trial. Nonetheless, inflammation can promote diseases, particularly late in life—largely beyond reproductive periods influenced by natural selection. Key for any successful intervention will be to dampen the smoldering inflammation of old age, while maintaining the critical roles of inflammation in recognizing and orchestrating responses to infections and damage.

Regular exercise has long been known to contribute to a substantially improved and longer health span, as well as to reduced cancer incidence (Mercken et al. 2012). Interestingly, exercise has been shown to increase autophagy, although it has not been demonstrated whether the physiological benefits of exercise require autophagy. Moreover, exercise reduces chronic inflammation, and we can therefore speculate that its ability to limit cancer incidence could be dependent on both increasing autophagy and decreasing inflammation. So exercise not only makes us more "fit" according to the popular meaning of this term, but it also permits our stem cells and tissues to maintain higher fitness, delaying aging and reducing selection for oncogenic events.

Purging Senescent Cells

A common argument from cancer biologists is that aging is the cost of cancer prevention—that telomere attrition and senescence in cells of old tissues are part of tumor suppressive mechanisms that necessarily also contribute to aging phenotypes (Serrano and Blasco 2007; Sharpless and DePinho 2007; Hoeijmakers 2009). While it is true that telomere shortening and cellular senescence do contribute to tissue dysfunction in old age, I argue that rather than preventing carcinogenesis, tissue dysfunction actually promotes it. While reduced cellular fitness associated with tissue dysfunction in old age may appear to represent the opposite of the high fitness cancer phenotype, the theory of adaptive oncogenesis helps us understand how reductions in stem cell fitness can promote selection for adaptive oncogenic events. The odds that a mutation improves fitness are much higher when a cell population is removed from a fitness peak.

Longer life spans among animals are correlated with shorter telomeres, typically in larger mammals that shut down telomerase (the enzyme that maintains telomere lengths) in adult tissues (Gomes et al. 2011). These evolved programs to limit telomere maintenance in adults are likely to contribute to tumor suppression in longer-lived and larger mammals by limiting the number of divisions that a cell can undertake. However, suppressing telomere extension in adults cannot explain longer lives, with concomitant maintenance of youthful and tumor suppressive tissue fitness landscapes. Telomere shortening leading to cellular senescence should instead promote tissue decline, indicating that other tissue maintenance programs contribute to delayed physiological aging and cancer for the evolution of long-lived mammals.

Recent elegant studies from Jan van Deursen and colleagues demonstrated that cellular senescence does indeed contribute to aging. Using mice engineered to delete senescent cells (with a simple chemical treatment), they showed that eliminating senescent cells starting in late middle age can prevent multiple aging phenotypes, including deterioration of the heart and kidney (D. Baker et al. 2016). Importantly, the mice with deleted senescent cells actually lived longer on average. The maximum life span was not significantly extended, indicating that there are other limitations to longer life. Additional studies from Jianhui Chang, Daohong Zhao and colleagues showed that deleting senescent cells could restore stem cell function impaired either by prior radiation exposure or by aging (Chang et al. 2015). Old stem cells were rejuvenated!

The success of this deletion strategy may seem surprising, given that senescent cells typically represent only a minor fraction of cells in a tissue, even for the very old. However, senescent cells are more than just poorly functional—they are detrimental to their neighbors. The lab of Judith Campisi has shown that senescent cells secrete a number of inflammatory cytokines (Coppe et al. 2010). As we have learned, inflammation can contribute to multiple age-related pathologies. The deletion strategy essentially removes cellular factories of inflammatory cytokines. Still, this strategy, like all strategies that interfere with evolved programs, likely comes with risks. Campisi's lab has shown that senescence induction is important for wound healing, as deleting senescent cells in a similar mouse model delays wound healing (Demaria et al. 2014). Most important for our topic, whether this strategy will reduce cancer development in old age is not known. A tissue with a more youthful phenotype would be expected to disfavor oncogenic adaptation. Moreover, since senescent cells secrete inflammatory factors that can contribute to cancer phenotypes (like invasiveness), eliminating these cells could limit later stages of tumor development.

However, the situation may be more complicated. Scott Lowe's lab has shown that senescent cells secrete cytokines that attract immune cells, leading to their clearance (Xue et al. 2007). Immune cell attraction may play an important role in allowing the recognition and elimination of cancer cells that survive therapies. Chemotherapy cannot kill every cancer cell (which can number up to a trillion), and the immune system, perhaps alerted by senescent cells, could help cleanse remaining tumor cells. Even during normal oncogenesis, a certain level of senescence could be induced, given the stresses experienced by these transformed cells. This alert system could be critical for keeping cancer rare (Kang et al. 2011). Thus, more studies will be needed to know whether a strategy of clearing senescent cells could lead to negative or positive changes in cancer susceptibility.

An Ounce of Prevention

While it would be wonderful to find a solution to the physiological decline and increased cancer risk associated with aging that involved simply taking a pill, care should be taken to not counter millions of years of evolution. The programs discussed above, including IGF1 signaling, autophagy, inflammation, and cell senescence, evolved for a reason. Interfering with these programs could counteract their critical roles in repair and defense. In addition, the targets of antiaging drugs, such as mTOR and autophagy, have pleiotropic effects, affecting many cellular pathways. Caution is advised when modulating key players, particularly when we do not yet understand all of the downstream effects to be expected. Still, since aging itself is largely due to the waning of natural selection's action to prevent it, we might be able to intervene in a subtler fashion to tune programs back to more youthful settings. For example, if we can learn more about the nuances of age-associated changes in inflammatory profiles, perhaps we can move these profiles toward a more youthful phenotype, including responsiveness to damage and infection. Natural selection is very good at tuning. Can we be?

Specific recommendations are difficult, given that the best data are from animal studies. Experiments with lab animals have the advantage that they can be well controlled, with the primary differences between experimental groups being the parameter of interest, like the ratio of carbohydrates to protein in diet. However, lab animals are not humans and are not experiencing the real world. Human studies are supportive of the conclusions from the lab studies, but they suffer from the many other variables encountered when comparing people with diverse genetics and lifestyles. Not surprisingly, humans are bad at reporting information about themselves—we tend to report that we have healthier diets and lifestyles than is actually the case. While specific recommendations, such as those regarding protein content in our diet, will have to wait, in general we already know how to extend the life span and limit disease: eat a balanced healthy diet, including lots of fruits and vegetables; exercise regularly; and don't smoke or otherwise exposure yourself to carcinogens.

CHAPTER 13

Understanding Clinical Data from an Evolutionary Perspective

Thus far in this book, I have mostly discussed cancer from a dispassionate and academic perspective. Cancer has a large impact on our lives, whether by increasing our suffering and ending our lives or by taking the lives of friends and loved ones. Roughly 40 percent of people living in developed countries will develop cancer, and about half of these individuals will die of their disease. As I discuss in this chapter, we can use evolutionary theory, including adaptive oncogenesis, to better understand why and when we develop cancer, how to interpret clinical data related to malignancies, and how cancer develops within us. Evolutionary theory can provide a framework for uncovering the mechanisms underlying cancer progression within an individual, as well as cancer incidence across the human population. In the case of an individual's cancer, mutational adaptations to diverse microenvironments engender genetic heterogeneity within the tumor and in metastatic sites. Our evolved life history can serve as the foundation for integrating clinical data on age-related detection of oncogenic mutations and their relationship to cancer risk. Across the human population, we can appreciate how evolutionary pressures over human history have affected patterns of cancer incidence as humans experience new lifestyles and environments.

Only by appreciating the ultimate cause of cancers can we hope to more effectively limit the human impact of this disease.

Stem Cells and Cancer Origins

As discussed above, tissues are organized hierarchically, with a small number of stem cells maintaining a much larger number of short-term progenitor cells and an even larger number of mature cells that perform designated functions for the tissue. This hierarchical organization is tumor suppressive, as the small number of infrequently dividing stem cells limits the target size for oncogenic mutations, while mutations that occur in more committed cells will be more likely to be lost via differentiation. Nonetheless, cancers frequently initiate in stem cells, particularly if the microenvironmental conditions favor somatic evolution. Given that stem cells are maintained in a tissue for the lifetime of the organism, these cells can accumulate mutations over a lifetime (notably, cancers appear to also initiate in nonstem cells—stress can induce reprogramming of differentiated cells to stem-like cells, and oncogenic mutations can confer stem cell–like qualities to a more committed cell) (Wahl and Spike 2017).

Some cancers appear to maintain a similar hierarchical organization, with cancer stem cells (CSCs) responsible for the long-term maintenance of the cancer mass (Reya et al. 2001). These CSCs can represent a very minor fraction (for example, less than 1 percent) of the overall tumor mass. Essentially, these cells behave like stem cells, except that due to mutational changes these cells now function to maintain a cancer instead of a normal tissue. Therefore, cancers can preserve some aspects of the hierarchy present in the tissue from which they were derived, with CSCs giving rise to more differentiated tumor cells with shorter cellular life spans. These cell types can even express many of the same proteins as the originating tissue, remnants of its somatic evolutionary history. Nonetheless, some cancers do not appear to have a distinct CSC population, and instead all cells of the cancer appear capable of propagating the tumor mass over time.

Therapies that fail to target CSCs may eliminate the bulk of the tumor, but the remaining CSCs will quickly repopulate the cancer and lead to relapse. For this reason, current research is geared toward the development of therapies that best target the CSCs, to achieve more durable remissions for cancer patients. The difficulty is that the CSCs often appear refractory to therapies. In particular, they survive the standard chemotherapeutic regimens for cancer better than the rest of the tumor does. They also appear to better tolerate many of the newer targeted therapies—such as imatinib for chronic myeloid leukemia, discussed in Chapter 5. In essence, the same mechanisms sculpted by natural selection that prevent oncogenesis in normal stem cells, such as infrequent cell divisions and the pumping out of damaging chemicals, are now exploited by the CSCs to increase survival in the face of therapies. Current research seeks to identify vulnerabilities in the CSCs that are not present in normal stem cells, which is not an easy task since CSCs appear to rely on many of the same pathways.

Studies of acute myeloid leukemia, which is highly associated with old age, have shown that the oncogenic mutations that initiate the leukemia are present in hematopoietic stem cells (HSCs), yet these oncogene-containing HSCs are not capable of perpetuating the leukemia (Corces-Zimmerman and Majeti 2014). Instead, a separate population of CSCs, which contains additional oncogenic mutations, has the ability to propagate the leukemia. The leukemia appears to initiate in an HSC, but with subsequent oncogenic mutations creating a distinct CSC population. Of note, leukemias that reoccur after therapy sometimes contain the initiating mutations that were present in the HSC population, but they lack some of the additional oncogenic mutations previously found in the CSCs of the leukemia at diagnosis. Therefore, the oncogene-containing HSC population may serve as a therapy-resistant reservoir for the future occurrence of the leukemia.

The second principle of our theory describes how perturbed microenvironments favor selection for adaptive oncogenic mutations in stem cells. If we could better understand how CSCs rely on this altered microenvironment, corrupted both by the original context (such as aging) and by the growing cancer, we might be able to target these interactions. CSCs now reside on a new peak in an adaptive landscape, and their fitness is dependent on the microenvironment, just like the fitness of the normal stem cells was dependent on the normal environment. Therefore, targeting the cancer microenvironment could be a way to reduce the fitness of the CSCs or favor a more benign phenotype. I will discuss this further in Chapter 15.

Changes in Clonal Frequencies with Age

The evolution of multicellularity gave rise to somatic evolution. Stem cells' decisions to differentiate or remain a stem cell are both stochastic and dependent on external factors. As genetic and epigenetic mutations accumulate in stem cells, different stem cell clones can be distinguished by the changes that they bear. Sometimes these mutations are phenotypic (for example, they change the proclivity of the stem cell to divide or renew itself), and selection will act on these phenotypic alterations. Other times, the mutations do not change phenotype, and thus such clones are subjected only to drift. With both selection and drift, the representation of the various stem cell clonal lineages in the pool will change over time.

As postulated in the theory of adaptive oncogenesis, evolution over millions of years led to stem cells that are well adapted to their respective niches, to maximize function while minimizing risks for malignancy. At the same time, somatic evolution acts within each individual, with selection for stem cell clones with the highest fitness. At least during youth, the interests of the individual coincide nicely with those of the stem cell, with the evolved plan for stem cell maintenance (typically, low division rates with efficient self-renewal) producing near-maximal somatic cell fitness. For this reason, oncogenic mutations typically reduce self-renewal, leading to clonal loss through differentiation (DeGregori 2012). This relationship changes in old age, for which germline selection for continued tissue maintenance has been weak or nonexistent, with oncogenic adaptation favored in aged tissue microenvironments.

Previously, it took dozens of researchers in multiple countries a decade to sequence one human genome (sequencing was first reported in 2000). In 2016, this sequencing can be achieved in a few days using one machine run by one person. This improved sequencing technology has allowed us to sequence thousands of human cancers, and many entire genomes (3×10^9 bases per haploid genome) have been sequenced—which has provided new insight into the underlying mutations. In addition, more recent studies have sequenced DNA from noncancerous tissues, which could provide insights into the earliest stages of oncogenesis.

When we sequence somatic cells, whether from a tumor or normal tissue, we are limited by our ability to detect variants. The DNA we inherit from our parents is our germline DNA, but this original code will accumulate mutations in somatic cells throughout our lives. Detection of these somatic mutations can be difficult. For a cancer, which is initially derived from a single cell, any somatic mutation that was in the initiating cell will be present in every cell in the tumor and will thus be detected at a high frequency. This mutation is termed a variant, and the variant allele frequency (VAF) refers to the fraction of genetic alleles from a particular genomic location that possess this variant. For a mutation present in one gene allele (one of the two copies, called heterozygous) in all cells of a tumor, the VAF should be 50 percent. Additional mutations will accumulate as the tumor evolves and will reside in only a fraction of the cancer cells. Many of these mutations will be present in the final cancer with much less than 50 percent VAF, particularly if they occur late and/or are not adaptive. We will more easily detect mutations that are advantageous (as they will be positively selected and thus clonally expand), as well as any other mutations that happened to be in the same cell as one of these driver mutations. These other mutations that come along for the ride are called, not surprisingly, passenger mutations. Thus, when we sequence a cancer, we gain insight into some but not all of the mutations present, and some but not all of these mutations can be oncogenic.

Depending on how deep we sequence—an expression that refers to the number of times we "read" a particular sequence—we have different detection limits, which are typically about 5 percent. While a tumor will have various clonal expansions for which associated mutations can be detected, normal tissues that are derived from many stem cells will have far fewer detectable mutations. Importantly, just because we do not detect mutations does not mean that they are not there. Imagine that we have 1,000 cells, each of which has a very small genome of only 1,000 bases. Each cell has one variant relative to the rest of the population, and initially no two cells have the same variant—every possible base is mutated across the population. If our sequence detection limit is 5 percent, or even 0.5 percent, we will only "see" the unmutated consensus sequence. Nevertheless, if one of these mutated cells turned out to be more fit than the other 999, and it expanded in the population to a frequency of 20 percent, we would now easily detect this mutation. We need to remember our limits of detection when considering somatic mutations and how they are acted on by selection.

Multiple recent studies leveraged next-generation DNA sequencing to detect somatic mutations in peripheral blood cells across thousands of cancer-free humans of various adult ages (Jaiswal et al. 2014; Genovese et al. 2014; Xie et al. 2014; McKerrell et al. 2015; Busque et al. 2012). A neutral mutation that was present in only one HSC would contribute to less than 1/10,000th of the peripheral blood cell pool. Since the detection limit in these studies was about 5 percent, which could detect a heterozygous mutation present in 10 percent of cells, mutations that had not been subject to strong positive selection would not be detected. Mutations that happened early in fetal development, or that "got lucky" by means of drift, could represent exceptions that surpass this threshold without positive selection. Importantly, mutations that were in cell clones clonally expanded to 10 percent or more of the pool would be detected, and such detected mutations could either be drivers (increasing cell fitness) or passengers. Strikingly, these researchers observed that clonal expansions exceeding 10 percent of cells were very rare in people under forty. In contrast, these expansions were quite common in the elderly, detectable in up to 20 percent of individuals over seventy. Most but not all clones possessed oncogenic driver mutations. Of relevance, clones typically contained only a single oncogenic mutation, indicating that multiple driver mutations are not necessary for the age-dependent pattern of expansion.

There are three possible explanations for this clonal hematopoiesis of the elderly: mutations are largely restricted to ages past forty (the most widely proffered interpretation); the clonal expansions may take a long time to achieve detectable abundance, in that the mutations could confer very modest fitness benefit; and particular mutations confer very different fitness effects on hematopoietic stem and progenitor cells in young and old individuals (consistent with the theory of adaptive oncogenesis). As over half of mutations accumulate in the hematopoietic system during our development to maturity, in line with higher division rates to generate adult-sized pools, we can reject the first explanation. While the second explanation cannot be ruled out for all mutations, I have discussed the direct evidence for how oncogenes can be strongly positively selected in old hematopoietic microenvironments while not being selected for in young microenvironments, which supports the third explanation (Henry et al. 2015). When aging is understood as a process that changes selective pressures, we can better appreciate how changes in mutation frequencies (clonal representation) with age can reflect alterations in selective pressures, as opposed to simply the generation of these mutations.

Oncogenic Mutations in Normal Tissues

Technological advances in the past few decades, including the sequencing technology discussed in the previous section, have been leveraged to identify mutations in multiple tissues. For example, BCR-ABL translocations are found in the blood of about one of three adults, yet the leukemia associated with this event, chronic myeloid leukemia, has a lifetime incidence of one in five hundred (Matioli 2002). Mutation or epigenetic silencing of common tumor suppressors, like PTEN and INK4A, are detectable in histologically normal endometria and breasts (respectively) of cancer-free women at rates that far outpace the incidence of the cancers associated with these mutations (Crawford et al. 2004; Mutter et al. 2001). Strikingly, cell clones with mutations that inactivate the p53 tumor suppressor gene are detectable, with about fifty clones per square centimeter of skin—and these clones are much bigger in skin taken from sun-exposed parts of the body (Jonason et al. 1996). P53 mutations provide protection from apoptosis of skin cells induced by exposure to ultraviolet (UV) light, which could explain why sun exposure promotes the clonal expansion of skin stem cells that possess a p53 mutation. Finally, a recent sequencing study of skin from eyelids identified thousands of mutations within about a square centimeter of skin, with up to fifty being within known oncogenes (Martincorena et al. 2015). How is it that we can experience so many oncogenic mutations, yet only about 40 percent of us will develop a cancer, largely late in life, and such cancers arise from a single mutated cell? Clearly, evolved tumor suppressive mechanisms are largely doing their job.

First, hierarchical tissue organization could be key, as oncogenic mutations that occur in nonstem cells (particularly mature cells) may not engender substantial risk of further cancer progression. In addition, the development of a cancer requires multiple mutations, and many of the mutant clones simply will not acquire the full range of oncogenic mutations required to generate an aggressive cancer. Extrinsic tumor suppressive mechanisms, such as immune surveillance and the normalizing abilities of tissue architecture, could further suppress cancer progression. Finally, in line with the theory of adaptive oncogenesis, these oncogenic mutations, particularly when they occur in healthy young tissues, may frequently be nonadaptive or minimally adaptive, reducing the fitness of stem and progenitor cells—for example, by inducing apoptosis or differentiation. Consequently, oncogenically initiated cells frequently fail to clonally expand to a point that would favor further malignant progression.

Computational Modeling of Somatic Evolution

Mathematical modeling supports the views that evolved tissue microenvironments play a key role in limiting the clonal expansion of oncogenically initiated cells, and that age-dependent changes in tissue microenvironments are important in promoting oncogenesis.

Andrii Rozhok and colleagues (2014) developed a model of somatic evolution in human HSCs. The model for the first time incorporated experimental data regarding HSC division rates and mutation accumulation, the bone marrow environment, and how these parameters change with age. The model is stochastic, in that chance plays a role in the impact of mutations on fitness and their representation within cell populations. Performing simulations in this model allowed researchers to observe how mutations in HSCs could affect their fitness status, leading to either the loss or expansion of particular mutationally marked clones over human lifetimes. These changes in genetic allele frequencies are referred to as clonal dynamics. Simulations showed that mutations, although necessary, cannot promote leukemia development without the age-dependent alterations in the tissue microenvironment that are conducive to the expansion of mutant cell clones. The decline in old age in the quality of the tissue microenvironment away from the evolved phenotype led to increased opportunities for the selection of adaptive mutations. Only when the influence of the microenvironment on cell fitness was incorporated could the model generate curves of clonal expansions within human lifetimes that accurately replicated the observed incidence of human leukemia. Modeling thus supported the key principles of adaptive oncogenesis.

The model provided a mathematical basis for the link between stem cell clonal dynamics and cancer risk, and it suggested that clonal hematopoiesis should be linked with higher risks of leukemia. Notably, this computational study was published concurrently with reports from multiple groups of researchers showing experimentally that clonal hematopoiesis increases in the elderly, as predicted by the model, and is indeed linked with higher leukemia risk (Jaiswal et al. 2014; Genovese et al. 2014; Xie et al. 2014; McKerrell et al. 2015). The mathematical modeling complemented the genomic studies of human blood cells, providing a mechanism—alterations in selection mediated by the decline in the microenvironment—to explain real-life results.

The evolutionary theory underlying this model can be traced back to Carl Sprengel and Justus von Liebig in the early 1800s. Sprengel showed how one environmental factor could limit organismal fitness, as when the growth of a plant is limited by the supply of nitrogen. Liebig expanded and popularized this concept, known as the Sprengel-Liebig's, or simply Liebig's, law of the minimum (Gorban et al. 2011). Liebig used the example of a barrel: removing the upper part of one vertical plank would limit the amount of water that the barrel could hold, independent of the status of the other planks.

In the 1930s, theoretical work by Victor Shelford described how every factor in the environment has an optimum intensity that a species (or a phenotype) is best adapted to, with the relationship between fitness and the intensity of the environmental factor conforming to a typical bell-shape curve. This is known as Shelford's law of tolerance, according to which an optimum intensity for a given environmental factor confers the highest fitness of the phenotype, with changes in factor levels in either direction decreasing the phenotype's fitness (Shelford 1931). Combined, Sprengel-Liebig's and Shelford's laws demonstrate that changes in the environment have an impact on the effects of mutant phenotypes on fitness, and that the selective value of different mutations may depend on very different environmental factors.

Following this theoretical platform, modeling and theoretical studies by Rozhok and colleagues showed that to understand somatic evolution, one must consider how tissue microenvironment determines the fitness impact of mutations. It is typically assumed that oncogenic mutations have defined positive impacts on cells' fitness (see Bozic et al. 2010). However, the effects of a mutation on fitness are not static and are highly dependent on the environmental context. In terms of the changes at the tissue level, we do not know which factors important for stem cells become limiting for stem cell fitness in aged tissues. Still, it is clear that alteration of microenvironmental factors away from the optimum should reduce the fitness of somatic cells, promoting selection for mutational changes that restore fitness. Even though not all mutations adaptive to the age-altered fitness landscape will contribute to cancer, some can. Conditions conducive to somatic evolution increase the odds that such evolution will lead to cancer.

Tumor Heterogeneity and Metastasis

While many studies have cataloged genes that are mutated in cancers—including The Cancer Genome Atlas project to sequence thousands of cancers—there is not one genotype for a given cancer. Within a cancer, there is substantial variation in genetics, with different clones bearing different mutations. In one of the most compelling demonstrations of tumor heterogeneity, Marco Gerlinger, Charles Swanton, and colleagues sequenced the genomes of distinct regions of renal cell carcinomas (kidney cancers) from different individuals (Gerlinger et al. 2014). Their analyses showed that different parts of the cancer in the same individual possessed distinct complements of driver genes. Only one driving oncogenic mutation was consistently shared across different regions of a person's cancer—a mutation in the VHL tumor suppressor gene. The VHL mutation was in the trunk of an evolutionary tree for each cancer, and major branches of the tree exhibited different oncogenic mutations. Metastases from the cancer were derived from particular regions of the parent tumor, and these possessed unique mutations as well.

By the time a cancer reaches a large enough size to threaten the health of the individual, it has accumulated a large number of mutations, from single base (letter) changes to larger rearrangements of chromosomes and epigenetic changes that affect how DNA is read. Both the large number of cell divisions required to generate this mass and the higher mutation rates that can characterize some cancers contribute to genetic heterogeneity in the cancer (Marusyk, Almendro, and Polyak 2012). Most of this heterogeneity consists of genetic changes that do not influence phenotype. The frequencies of these mutations in the cancer population are dictated by drift and by hitchhiking with phenotypic mutations that do affect cell fitness. Other mutations affect phenotype, and some of them can contribute to the hallmarks that render cancers so dangerous. Additional phenotypic heterogeneity can be nongenetic, resulting from the hierarchical organization that persists from the original tissue structure, with more stem cell–like cells and other cells of variable differentiation status.

If genetically driven phenotypic diversity were generated in a fully intermixing population of cells in a stable and uniform environment, then selection would act to favor the genotype that optimizes fitness, with movement of the population up a fitness peak. While other peaks could be explored, eventually only the cells that occupied the highest peak would prevail. Phenotypic diversity in the cell population would be minimized. However, advanced cancers actually experience very unstable and variable environments, and cells in different regions of the same solid cancer are unlikely to be in competition with each other. Thus, different fitness landscapes in the same tumor lead to selection for diverse phenotypes.

Cancers can reside in very nasty neighborhoods, with high acidity, low oxygen, and inflammation (Marusyk, Almendro, and Polyak 2012). The degradation of the microenvironment initially results from the underlying cancer-promoting context, like smoking, but is amplified and further altered by the growing tumor itself. Just as a healthy and stable microenvironment promotes the status quo (according to the theory of adaptive oncogenesis), a degraded and unstable microenvironment stimulates further change by selecting for adaptive phenotypes. The microenvironment of the cancer can be highly variable in different regions. For instance, the inside parts of tumors are often poorly oxygenated (hypoxic) relative to the periphery, due to inadequate blood supply—which leads to selection of phenotypes adaptive to hypoxia. In addition, the periphery of the tumor should experience more contact with non-tumor cells of the host, including immune cells, normal tissue, and vasculature—which creates distinct selective hurdles. Notably, microenvironmental variability is not just spatial, it is also temporal. For example, poor vascularization of tumors can lead to changes in oxygen and nutrient supply to a given area of the tumor over time. Heterogeneity also extends to noncancer cell types. The tumor mass can be up to half nontumor cells and can include fibroblasts, blood vessels, macrophages, and other immune cells.

The heterogeneous tumor landscape, both temporal and spatial, can select for the migratory ability of cancer cells (Daoust et al. 2013). The propensity to migrate should be disadvantageous in a stable, rich environment, where resources are better spent on cell division than on migration. However, being migratory becomes advantageous in heterogeneous and unstable environments, where "greener pastures" can be sought. These ecological concepts have been well described for animals and their propensity to migrate. Computational modeling has demonstrated how microenvironmental heterogeneity selects for cancer cells with invasive and migratory phenotypes (Chen et al. 2011; A. Anderson et al. 2006). Given that cancer cell migration is a key step in the spreading of a cancer to other organs (metastasis), it is not surprising that tumor heterogeneity is associated with poor outcomes for cancer patients. In addition to selecting for more aggressive phenotypes, variable tumor microenvironments also maintain the phenotypic diversity from which drug-resistant cancer cells can be selected upon therapy.

Primary tumors rarely lead to the death of the individual. It is the metastatic spread of the cancer to other organs that typically results in the death of the patient, usually by disrupting the functions of critical organs such as the lungs, liver, brain, and bone marrow. Metastasis involves cancer cells from the original tumor moving through the surrounding matrix, entering a blood or lymph vessel, surviving there, avoiding immune elimination, leaving the vessel for a new tissue, and establishing themselves within an entirely different tissue niche (Nguyen, Bos, and Massague 2009). Clearly, metastasis requires that the cell clone overcome multiple substantial hurdles. Of the millions of cancer cells released into the bloodstream, only a handful successfully establish metastatic clones. Nonetheless, for many cancers, metastatic spread will be inevitable if they are left untreated. The size of the primary tumor matters, as a larger tumor will generally release more cells—which increases the odds that a few cells will be successful at distant sites. Importantly, microenvironmental hurdles and heterogeneity in the parent tumor facilitate selection for phenotypes that aid metastasis, such as the migratory phenotype described above. Thus, harsher and more variable microenvironments of the parent tumor select for more aggressive clones that are better able to thrive in the primary tumor, but that consequently are endowed with the traits that facilitate metastatic spread.

While some cells can be migratory, there are limitations to the extent of intratumor competition, in part dictated by spatial separation of tumor cells. Cells on one side of a solid tumor ten centimeters wide are in limited competition, if at all, with cells on the other side. This fact can in part explain the genetic and phenotypic divergence evident in different parts of the same tumor. These different regions will present different microenvironments, and the independent evolution of cancer cells in the different regions is reminiscent of allopatric speciation, in which geographical isolation of a population facilitates the evolution of a new species (Mayr 1982). Similarly, the restriction (even if not absolute) of competition between different parts of a tumor can lead to different clones dominating different regions of the tumor (Lipinski et al. 2016). Indeed, the spatial restriction of solid tumors may permit the persistence of clones with lower fitness, including those with genomic instability—usually a disadvantageous state due to the generation of detrimental mutations. This persistence could then increase the chances that a clone adaptive to the tumor's microenvironment will emerge.

Surprisingly, even leukemias—which we think of as "liquid cancers" that are freely intermixing—exhibit substantial genetic heterogeneity, with distinct clones with different oncogenic driver mutations coexisting in the same patient (K. Anderson et al. 2011). Perhaps insufficient time has passed for one genotype to sweep through the population, or perhaps these different clones are differentially adapted to separate niches in the patient, serving as harbors for genetic diversity. Regardless, it is clear that this clonal diversity can serve as a reservoir for therapeutic resistance, as a previously minor leukemic clone can dominate after therapy.

Cancer cells in different regions of the same tumor can follow relatively independent evolutionary trajectories. As strong evidence that common selective forces are present across renal cell carcinomas (RCCs), the genomic studies by Gerlinger and colleagues (2014) revealed frequent convergent evolution leading to similar phenotypes instituted by independent mutations, often in the same gene but via distinct sequence changes. For example, the SETD2 tumor suppressor gene independently mutated in multiple branches of the same RCC, and the same pattern was observed for RCCs from multiple patients. Thus, the landscapes in geographically separated regions of the tumor can maintain features in common, leading to selection for the same or a similar phenotype. It is clear that loss of SETD2 activity was strongly selected for independently in different regions of the same cancer, although we do not understand the common landscape feature that led to this selection.

Notably, similar studies of colon cancers by Rose Brannon, David Solit, Michael Berger and colleagues did not reveal such high driver mutation heterogeneity, and different regions of the same cancer exhibited the same set of known driver mutations (Brannon et al. 2014). The heterogeneity that was observed primarily affected genes that do not play known driver roles in cancer. Thus, colon cancers appear to have a large trunk that contains most driver mutations, and the branches may contain nonphenotypic mutations that are fixed in different cancer regions by chance and/or more minor driver mutations that have not been well characterized.

Given the importance of cancer heterogeneity for prognosis and predicting the development of resistance to treatment, methods to assess that heterogeneity are needed. When a tumor biopsy is taken, the region of the cancer sampled is relatively random, and only a fraction of the possibly many tumor genotypes are measured. Different parts of the same cancer could even exhibit divergent phenotypes, such as invasiveness. More recently developed techniques are analyzing circulating tumor DNA, which is isolated from tumor cells and vesicles that were released into the blood of the patient. Circulating tumor DNA appears to provide a better sampling of multiple parts of the cancer at once, and it may reveal more information about the diversity of genotypes and phenotypes in the cancer (Zhang et al. 2016). Armed with such information, physicians should be better able to design treatments and predict mechanisms for the evolution of treatment resistance.

Ancient Genomes in a Modern World

Cancer is clearly much more common in the twentieth and twenty-first centuries than it has ever been before. Since we are living longer, a greater fraction of the population is at the older ages where cancer incidence is higher: our successes in extending life have led to increases in diseases of old age like cancer. We are also altering our bodies, often by choice, in ways that heighten cancer risk—such as by smoking, being overweight, or creating polluted environments. An additional contributor to increased cancer rates derives from a factor that we have less control over: our genomes have been almost entirely sculpted by living conditions very distinct from those that we now experience. In some ways, we are maladapted to the modern world, at least in developed countries.

As one example, breast cancer is rare in contemporary traditional communities like the Dogon of Mali, but it is much more common in women in developed countries (Strassmann 1999). Like the Dogon, women in the ancient world spent most of their premenopausal lives either pregnant or nursing. Babies were breastfed for about three years. Pregnancy and nursing both suppress menstrual cycles, and these cycles are associated with high levels of estrogen and progesterone. These hormones may have promotional effects on breast cancer development, and less exposure to them could reduce cancer risk. It is estimated that women in ancient times experienced far fewer menstrual cycles than modern women, who reach sexual maturity earlier, often delay reproduction, have fewer children, and nurse these babies for less time. Even among modern women, earlier pregnancy, more pregnancies, and breast-feeding have been shown to be protective against breast cancer (Layde et al. 1989). Natural selection acted to favor a human genome for one type of lifestyle, in this case, frequent reproduction for women.

Evolution in humans is not fast enough to keep up with changing modern environments. The evolved phenotype for the breast was sculpted to match the lifestyles of women over many millennia. Modern lifestyles have led to alterations in the landscape of the breast, affecting stem cells and their niches—which, from the perspective of the theory of adaptive oncogenesis, should lead to selection for somatic cell adaptations mediated by oncogenic events. While we do not know the relative contributions of greater hormone exposures and altered adaptive landscapes, we can nonetheless appreciate how changes in lifestyles can affect cancer risk. Further exploration for how modern disconnects from our evolved genomes can contribute to cancer risk could suggest preventive interventions.

Women who inherit alleles of BRCA1 and BRCA2 tumor suppressor genes with mutations that cause a loss of function are at substantially increased risk of developing breast and ovarian cancer. While the women are born with one good allele of BRCA1 or BRCA2 and one defective one, the good copy of the BRCA gene is lost during tumor evolution. Male carriers of these defective BRCA alleles are also at increased risk of breast cancer (normally rare in men). The encoded BRCA proteins are important for DNA repair, among other roles. Interestingly, members of families bearing mutated BRCA1 or BRCA2 appear to conceive higher numbers of offspring, with shorter birth intervals (Kwiatkowski et al. 2015). This increased fertility is also evident in studies of familial lineages with BRCA mutations who conceived children before 1930 (before the common use of reliable birth control) (K. Smith et al. 2012). Female carriers of these alleles experience fewer miscarriages, and higher fertility is observed among male carriers. While these alleles increase cancer risk after age forty, they could have increased the fitness of carriers by increasing fertility—which may have contributed to the maintenance of these alleles in the population.

Human migrations can also influence cancer risk. People of Northern European ancestry have pale skin. To produce vitamin D, our bodies require exposure to UV light, and dark melanin pigment in our skin blocks this light from penetrating deep enough (Jablonski and Chaplin 2010; Greaves 2014). Melanin also provides protection from the mutagenic and cancer-promoting features of UV light and prevents depletion of folate (an essential B vitamin) mediated by UV light. This trade-off has led to selection for lighter skin color (less melanin produced by melanocytes in the skin) in northern latitudes, where sunlight is scarce and more of the body covered by clothing, and for darker skin color in more equatorial regions where sun exposure is high. In the habitats where each population evolved, the proper balance is achieved: sufficient penetration of UV light to process vitamin D, but not so much as to create an excessive risk of skin cancer or to deplete folate. Of course, human migration to other regions disrupts this evolved balance. When British people migrated to Australia, their rates of skin cancers like melanomas markedly increased. Their evolved adaptation to the overcast skies of the British Isles conflicted with the strong sun they were exposed to Down Under. The increased risk of skin cancers has been largely attributed to greater frequencies of mutations induced by UV light.

Through the lens of the theory of adaptive oncogenesis, I would hypothesize that the skin of people of British ancestry living in Australia, which is excessively damaged by UV light, no longer reflects the normal (evolved) microenvironment for their skin stem cells, which leads to selection for adaptive oncogenic mutations. We can imagine that the strategies humans evolved to minimize death from cancer through years of likely reproduction differed in disparate regions of the globe, dependent on distinct external pressures. When people move, not only are they at least somewhat maladaptive to the new environment, but tissues receiving new environmental challenges (like more exposure to UV light) can engender reduced adaptation by resident stem cells. Just as the early humans who migrated to Northern Europe adapted to reduced exposure to sunlight through reduced melanin expression, there will be selection for adaptive mutations within stem cells in these newly challenged tissues. Such alterations of selective pressures, combined with increased numbers of mutations induced by UV light, will enhance skin cancer risk.

In contrast, people of African descent living in the United States are more likely to experience vitamin D deficiency than those of European descent, which could contribute to the higher incidences of heart disease, diabetes, and cancer (all associated with vitamin D deficiency) in African Americans (Harris 2006). Thus, while darker skin was adaptive in Africa, in regions with less sun and given modern habits of covering our bodies with clothes, the failure to effectively use UV light to process vitamin D becomes detrimental. Evolved trade-offs have been disrupted, with the need for the adaptation (dark skin) lost in the new environment, but the cost of this adaptation amplified.

It is not just our genes that we need to consider. The roughly forty trillion cells in a human are paired with a similar if not greater number of microbial inhabitants, including on our skin, in our mouth, and most numerously in our large intestine (Rook and Dalgleish 2011). We have evolved with our gut microbiome over the millennia and perhaps even millions of years of hominid evolution. A healthy gut microbiome is essential for the synthesis of certain vitamins, proper digestion, water absorption, barrier function, and immune regulation. Due to changes in hygiene, pathogen exposures, diet, and lifestyles in developed societies, our gut flora has changed immensely. For example, the microbial intestinal flora of Western humans is very different from that of peoples of rural Africa (Ou et al. 2013). Colorectal cancer is associated with dysbiosis, an imbalance in gut bacterial species that contributes to chronic inflammation. Therefore, given lifestyle changes, modern humans in industrialized nations possess intestinal tracts and immune systems that are no longer adapted to their gut microbial flora, which leads to inflammatory bowel disease and cancer (Rook and Dalgleish 2011). I would argue that this state of maladaptation, including by intestinal stem cells, leads to selection for adaptive oncogenic mutations, thus increasing colon cancer risk.

The Masculinity Penalty

For males, the main evolutionary task—reproduction—simply involves having sex, without the burdens of pregnancy, childbirth, and nursing (and for many species, child care). However, there is a cost. If you are a peacock, attracting a female means displaying a very gaudy array of feathers, which are expensive in terms of their production and maintenance, and perhaps more importantly because they increase the risk of death by predation. Similar, albeit not so histrionic, patterns characterize many birds. For many animals, being male entails taking more risks, more fighting (particularly over female mates), and thus greater odds of dying early. Higher death rates for males are evident in human hunter-gatherer societies and throughout human history, as well as in our primate cousins (Alberts et al. 2014). Young men are more likely to die due to risky behavior, such as in car accidents. Independent of accidents, men physiologically age faster than women, and die at earlier ages than women (by an average of 5–7 years). As a result, in the United States women account for more than 60 percent of those alive past the age of eighty and more than 80 percent of those alive past a hundred (Centers for Disease Control and Prevention 2014).

Natural selection appears to have invested more in youthful robustness (size, strength, and other attributes of masculinity) for males to improve success in mating, which could contribute to late-life decline. The investment is front-loaded. In fact, at least one of the pathways that is known to contribute to animal size and youthful robustness is the IGF1 or mTOR pathway described in the preceding chapter, a pathway known to contribute to aging (Blagosklonny 2013). Mutants that down-regulate IGF1 or mTOR live longer, but with lower fertility and virility. Natural selection clearly favors youthful male robustness over longevity if it means higher odds of reproductive success. In contrast, the greater need for longer female life to care for offspring, including grandchildren, combined with reduced external hazards, may have contributed to longer somatic maintenance and thus longer potential life spans.

In this light, understanding cancer as determined by evolved life history may explain the earlier incidence of some cancers in males. Analyses of incidence curves for cancers (brain cancer, bowel cancer, acute myeloid leukemia, chronic myeloid leukemia, and chronic lymphocytic leukemia) that are neither sex-specific nor substantially due to smoking reveal earlier (leftward) incidence for males (National Cancer Institute 2014). Similar male-earlier patterns are observed for cancers of the kidney, pancreas, and bladder, but these are linked with smoking and thus could reflect increased tobacco use among males. The earlier physiological decline for males could elicit earlier oncogenesis relative to females.

Due to greater natural selection for youthful robustness over longevity in males, aging-related changes to tissue microenvironments (earlier relative to females) should lead to selection for adaptive mutations, and some of these mutations should be oncogenic. The resulting cellular expansions of oncogene-bearing clones would then increase the risk of subsequent progression to full-blown malignancy, as discussed above, with age-dependence that is leftward-shifted relative to females.

CHAPTER 14

The Causes of Cancer in Early Childhood

Any death from cancer is tragic. Cancer causes suffering as it destroys our tissues and organs. While cancer in a seventy-year-old causes pain both for the patient and his or her family, cancer in a seven-year-old or a seven-month-old is particularly heartbreaking. The greatest medical and societal advances of the twentieth century in human health were not in extending the maximal human life span, but in greatly eliminating deaths early in life. Huge reductions in childhood mortality in the first half of the century largely derived from improvements in sanitation, the development of vaccines and antibiotics, and advancements in hospital care. Progress was also made in treating cancer. Up until the 1960s, childhood cancers were almost invariable lethal, with survival for five years past diagnosis at less than 10 percent. Survival with these diseases was typically measured in months, not years. The development of chemotherapy regimens for the most common of childhood cancers, acute lymphoblastic leukemia (ALL), led to progressive improvements in outcomes. By 2010, this previously lethal leukemia was cured in more than 90 percent of children (Hunger and Mullighan 2015). For other childhood cancers, such as acute myeloid leukemia (AML) and brain cancers, progress has been much slower, and cure rates remain unacceptably low. Given that natural selection acts to maximize individuals' reproductive potential, childhood cancers appear to be an enigma. This chapter explores some of the different evolutionary explanations for childhood cancers, including disconnects between modern environments and those of our ancestors, recent human adaptations that may have carried costs in terms of cancer risk, and the role of chance in the much smaller stem cell pools of early life.

Limited but Not Eliminated by Natural Selection

I have discussed how natural selection maximizes reproductive success, and thus the selective pressure to prevent disease or death wanes as we get old—leading to maladies of old age such as heart disease and cancer. Looking through the impartial lens of evolutionary biology, I will explore why natural selection does not eliminate cancer risk for children, given the huge fitness cost.

First, we need to consider relative risks. About 1 in 500 children develops cancer before the age of fifteen, with about one-third of the afflicted children developing leukemia. The overall childhood risk of developing the most common of these leukemias, ALL, is about 1 in 2,000. For most of our evolutionary history this risk was much lower than other threats, such as infectious disease. Natural selection does not (and cannot) eliminate risk, only minimize it to the point that maximizes the return on any required investment. Natural selection is also limited in its resolution by the power of drift. As discussed above, the effective population size for humans, which has determined the relative strengths of selection and drift, is about 10,000. If a certain genotype results in an increased childhood cancer risk of 1 in 10,000 (0.01 percent) or less, purifying selection will have little power to eliminate it. It will persist in the human population by chance alone. The contribution of inherited genetics to the risk of childhood ALL is unknown, but is likely less than 25 percent (noting that 25 percent of the 1 in 2,000 overall risk is quite close to the 1 in 10,000 drift barrier) (Enciso-Mora et al. 2012). While maternal and childhood exposures to carcinogens (chemicals and radiation) and maternal diet have been associated with increased risk for childhood leukemias, these factors do not appear to account for the majority of childhood cancers.

The Costs of Adaptations

Cancers in children primarily originate in three systems: the brain, bones and hematopoietic system. Armand Leroi and colleagues speculated on why this is the case (Leroi, Koufopanou, and Burt 2003). Despite common misconceptions, humans are still evolving, and this evolution has accelerated in the past 10,000–1,000,000 years as hominids and humans have encountered new challenges. The size of the hominid brain has expanded in response to changes in the environment, including the advent of cooking that facilitated acquisition of the increased calories needed to fuel our brains (Wrangham and Carmody 2010). The demands of group living, including in terms of communication, also drove evolutionary changes in our brains. Leroi and colleagues proposed that brain cancer in children could represent a cost of these changes, with the risk somehow increased due to alterations in developmental programs. Similarly, our skeletons have undergone significant remodeling in recent evolutionary history, and the authors speculate that these changes may have somehow increased the odds of cancer. Finally, we have experienced new pathogenic threats in the continuous arms race with microbes and worms. These threats increased in the past 10,000 years due to the development of agriculture and animal domestication. Viruses and other agents have made the jump from domesticated animals to humans many times, and the increases in human population density due to agriculture allowed new pathogens to thrive among us. The rapid evolution of the immune system to keep up with these new challenges may have come with the cost of increased leukemia risk. In each case, if childhood cancers were a cost of these brain, bone, or immune adaptations, the cost was clearly less than the benefit, given the rarity of the cancers. Notably, for each of these examples, we lack a mechanism to explain these potential trade-offs.

Large Risk from Rare Genetic Alleles

Genetic inheritance accounts for a small fraction of childhood leukemias, with most of this risk coming from common but weak alleles that only modestly influence risk (Enciso-Mora et al. 2012). Rare but strong inherited genetic alleles can confer a much larger risk of leukemia development.

Genetic risk factors for childhood leukemias include rare disorders like Fanconi's anemia (FA), ataxia-telangiectasia and Down syndrome (Hemminki and Jiang 2002). FA is a familial disease resulting from inheritance of loss-of-function alleles of one of at least thirteen different FA genes that encode proteins important in a DNA repair pathway. Individuals with this disorder exhibit severe anemia due to very poor blood cell production in the bone marrow, with hematopoietic stem cell (HSC) impairment at the foundation of this hematopoietic collapse. Impaired hematopoiesis is a common feature of a number of inherited disorders that result in debilitated DNA repair.

Individuals with FA are also at a 700-fold increased risk of developing myeloid leukemia, which primarily afflicts children, as well as other cancers. While the predominant interpretation for the link between the inheritance disorder and leukemia has been that disrupted DNA repair fuels leukemia genesis through mutation generation, Grover Bagby and colleagues provided a different interpretation (Bagby and Fleischman 2011; Lensch et al. 1999). They proposed that hematopoietic stem and progenitor cells from individuals with FA are defective in survival, which leads to strong selection for mutations that provide resistance to this enhanced cell death. Indeed, Bagby and collaborators showed that FA hematopoietic cells are highly sensitive to particular inflammatory cytokines, including tumor necrosis factor-α (TNF-α), but that myeloid leukemias that develop in individuals with FA are resistant to the same challenges. The researchers directly demonstrated that by inhibiting the proliferation of FA progenitors, TNF-α promotes selection for transformed clones resistant to this cytokine, facilitating the first step in oncogenesis (Li et al. 2007).

The data and explanations from Bagby and colleagues are consistent with the general principles of adaptive oncogenesis. We can see how inheritance of mutant FA genes greatly increases leukemia risk by three means: increased frequency of mutations, due to impaired DNA repair; inflammation promoted by DNA damage, which in turn creates a microenvironment that is particularly debilitating to FA hematopoietic progenitors; and cytokine-mediated impairment of these FA progenitors, which engenders strong selective pressure for adaptive mutations—some of which are oncogenic. A clear demonstration that increased frequency of mutations, while contributing to leukemia genesis, is insufficient by itself comes from FA patients who exhibit spontaneous reversion of the FA mutation in an HSC, which reverses both the bone marrow failure and the leukemia risk (Bagby and Meyers 2007). Adaptation mediated by somatic mutation need not be oncogenic, since the defect can be corrected nononcogenically. The corrected HSC can outcompete FA mutant HSC, thus abrogating the selective pressure for oncogenic adaptation by restoring healthy HSC competitiveness and mitigating leukemia risk despite all of the mutations that have occurred in FA cells.

Delayed Exposures to Infections

The clustering of childhood leukemia cases in particular cities in the United Kingdom and United States, while initially ascribed to industrial or nuclear exposures, was later shown to be more likely explained by exposures to other people. Leo Kinlen proposed that higher childhood leukemia rates could result from population mixing following influxes of newcomers, who introduced new pathogens into nonimmune populations (1988). He ascribed the increased risk to direct action of the pathogen on leukemia generation. Concurrently, a different but overlapping model was proposed by Mel Greaves (2006)—that newborns in developed societies experience much less exposure to microbes, some pathogenic and some not. While more hygienic conditions have led to enormous societal benefits in terms of reduced infant mortality, Greaves proposed that subsequent infectious exposures in later years (three to five years of age), such as when a child enters preschool, can lead to excessive immune activation. This hyperactivation of the immune system, with associated inflammation, could then be promotional for leukemias, perhaps by increasing the rates of mutations in hematopoietic progenitor cells or by promoting selection for cells with particular oncogenic events. There are clear parallels with the hygiene hypothesis for childhood allergies and type 1 diabetes, for which the increased incidence in developed nations likely results from insufficient immune stimulation in early life leading to overreaction later in life. In fact, around the world allergy rates and type I diabetes and ALL incidence are all concordant geographically. Our immune systems have been tuned by evolution to the conditions that were present for the vast majority of this evolution, which included ubiquitous exposure to pathogens and other microbes in early life.

So what is the evidence for the delayed infection model? The risk of acute leukemias for children under five increased several-fold in the twentieth century in developed Western nations, and subsequently in Japan and China (Greaves 2006). The risk of childhood acute leukemias is several-fold lower in India and the Caribbean, and perhaps even tenfold lower in black South Africans. Children who are in playgroups or day care in the first year of life, which presumably precipitates early exposures to infectious agents, have about twofold lower leukemia risk. Moreover, having an older sibling, with the associated increases in exposures, has also been shown to lower risk. Finally, genetic alleles for certain immune genes are associated with ALL risk, and these alleles may be involved in more potent immune responses. Greaves has speculated that such hyper-responsiveness may have been advantageous under more ancestral con-texts of high infectious risk to health and survival, which still exists in many of the underdeveloped parts of the world. It is only under modern conditions in developed nations that this hyperimmune responsiveness comes with the cost of increased leukemia risk.

In Greaves's delayed infection model, initiating mutations—which frequently represent chromosomal translocations that generate a novel oncogenic fusion gene—occur during fetal development or very early in life, followed by some subsequent secondary oncogenic mutations and promotional events. The detection of the oncogene-generating chromo-somal translocations at birth is typically about a hundredfold more common than the leukemia associated with these events. The question then is: how does hyperimmune stimulation promote leukemias? One potential mechanism is by selecting for progenitor clones that bear common translocations. The most common translocation in B-cell lin-eage ALL is ETV6-RUNX1, and studies have shown that B-cell progeni-tors bearing this oncogenic fusion are resistant to a cytokine stimulus associated with dysregulated immune responses, transforming growth factor-β (TGF-β)—which kills normal progenitors without this oncogene (Ford et al. 2009). Such an increase in selection for an oncogenic event, mediated by a change in the microenvironment, can be considered as analogous to the age-dependent alterations in oncogenic selection proposed by the theory of adaptive oncogenesis. Interestingly, this oncogene actually slows the expansion of B-cell progenitors in the absence of TGF-β, which could indicate a fitness disadvantage conferred by this oncogenic fusion under more normal conditions (in line with stabilizing selection to maintain the evolved phenotype). Thus, should an infant be born with ETV6-RUNX1-expressing stem and progenitor cells, and then a few years later experience an infection that leads to hyperimmune activation, these ETV6-RUNX1 cells would be at a distinct advantage due to preferential elimination of normal progenitors. This cytokine-driven positive selection could then lead to clonal expansion of the ETV6-RUNX1 cells, increasing the odds of further leukemic progression.

Bad Luck and Childhood Leukemias

If cancer incidence simply reflects the time-dependent accumulation of oncogenic mutations in stem cells, then the greater frequency of some cancers such as leukemias in children under the age of five than in older children or young adults is puzzling. Since stem cells should accumulate more mutations with age, what could account for the increased incidence of childhood cancers in young children? When Andrii Rozhok designed the computational model of somatic evolution described in the previous chapter, he aimed to characterize the possible parameters controlling this evolution of HSCs in adults with age. It was thus surprising when the model generated increased clonal expansions in very young children, recapitulating the increased cancer risk observed at these same ages (Rozhok, Salstrom, and DeGregori 2016). The model Rozhok developed uses the Monte Carlo method, which runs repeated simulations to estimate probabilities of outcomes given a range of variables. The beauty of a Monte Carlo model is that results essentially emerge from a black box, dependent on the data entered. Importantly, the evolutionary process during simulation is not under the control of the programmer. Since Rozhok was very careful to use experimentally generated data for all parameters where such data were known, including how these parameters change from birth until very old age, the model was able to generate a potential solution to the childhood leukemia conundrum even though he and his colleagues had not asked the relevant question up front. Changes in stem cell pool size and division rates turned out to be key determinants of somatic evolution and cancer risk (Rozhok, Salstrom, and DeGregori 2016).

During fetal development, the first of the stem cells that will produce blood cells for your whole life, called definitive HSCs, are born from the endothelial (blood vessel lineage) layer of the fetal aorta. These endothelial cells are thought to change into HSCs and then be expelled from the endothelial layer. Very few of these cells, on the order of a few dozen or less, are thought to actually contribute to long-term hematopoiesis (Catlin et al. 2011). By the time humans are born, there are only a limited number of HSCs maintaining our blood production, roughly proportional to the smaller size of our bodies at birth relative to adulthood (Abkowitz et al. 2002).

Exploring the parameters of his computational model, Rozhok showed that the small size of the HSC pool at and before birth makes it possible for drift to influence the genetic makeup of the pool over time (the clonal dynamics). In fact, even if the fitness effects of accumulating mutations were completely turned off, the model still generated the early childhood increase in clonal expansions. The increased incidence of clonal expansions early in childhood could be driven by drift alone, with selection playing a secondary role in determining the size of these expansions. The smaller Rozhok made the pool, which should increase the power of drift, the larger the fraction of the HSC population was occupied by a single clone during the first few years of life in the model.

The other key factor was HSC division rates. As mentioned previously, HSCs in adults divide only a little more than once per year, which is sufficient to maintain the HSC pool. However, this adult-size pool must be generated in the first place, which requires a large expansion of HSCs from the few generated in the fetus to the adult number. Therefore, division rates are at least thirty times higher in early childhood. This higher cell cycling rate operates synergistically with the drift-driven expansion of HSC-bearing oncogenic mutations: the clonal expansion would not only be larger but would be dividing much faster, increasing the odds of sequential mutation accumulation and oncogenic progression. Cancer development requires the acquisition of multiple oncogenic mutations within single cell clones. As discussed in the previous chapter, oncogenic clonal expansions engender cancer risk, as the target size for the next mutation in cancer development is larger. Higher cell cycling also increases the risk, as mutations are generated during DNA replication.

Further manipulation of the model showed that drift did not significantly influence clonal evolution in adult HSC pools, which are much larger. Instead, the fitness value of mutations was key, and as shown in the previous chapter, increased HSC clonal expansions in old age mirroring known leukemia rates were generated only when the fitness effects of mutations were tied to microenvironmental change, consistent with the theory of adaptive oncogenesis. During young adulthood, the low levels of clonal expansions and consequent low leukemia risk could be explained by the strong stabilizing selection present in youthful microenvironments: most mutations that changed phenotype reduced fitness, and these clones were eliminated. In all, these computational studies showed that leukemia risk in life is governed by different evolutionary forces during different phases of life (Figure 14.1), with drift-driven oncogenic clonal expansions early in life, stabilizing selection preventing clonal expansions through years of likely reproduction, and positive selection promoting the expansion of clones bearing oncogenic mutations adaptive to the aged microenvironment late in life.

No Simple Answers

There are times when there are multiple competing models to explain some scientific or medical conundrum, for which only one model can be correct. For the explanations for childhood cancers discussed above, it is unlikely that a single explanation will have such primacy: the explanations are not mutually exclusive. We can hypothesize how the roles of drift, infection, and genetics could interact. The drift-driven expansion of HSC clones with an initial oncogenic mutation should be suppressed during childhood as the stem cell pool expands, which restores the power of selection to eliminate oncogenic clones. However, this period after birth—during which the oncogene-initiated clone exists—creates risk, in that the odds of the next oncogenic event (which could negate the cost of the first event) will be proportional to the size of the clone and the time that it persists. This period will also create opportunity for

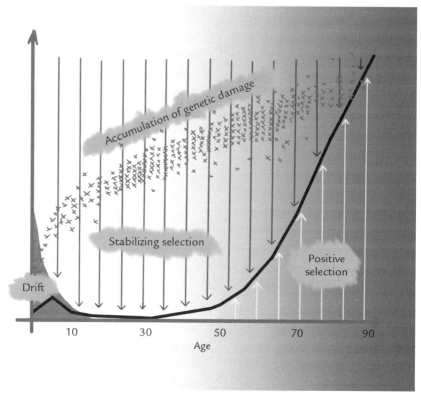

Figure 14.1. Changing roles of somatic selection in life. Mutations accumulate rapidly in hematopoietic cells early in life, coinciding with rapid body growth, and then more slowly after maturation. In early childhood, the small size of the hematopoietic stem cell (HSC) pool increases the role of drift in the fate of mutations, which facilitates the expansion of oncogene-bearing cell clones. Through periods of likely reproduction, stabilizing selection is dominant in the larger HSC pools, which leads to suppression of oncogenic clonal expansions. Late in life, the degradation of the bone marrow microenvironment leads to positive selection for adaptive oncogenic mutations, which in turn leads to greater risk of oncogenic clonal expansions. The expansion of HSC clones bearing oncogenes increases the risk of leukemia. The mutation accumulation data is based on epigenetic changes in life in blood cells from Horvath, S., "DNA Methylation Age of Human Tissues and Cell Types," 2013; the pattern is similar for DNA mutations in mouse hematopoietic cells (Vijg, J., et al., "Aging and Genome Maintenance," 2005). The leukemia curve is graphed from data tabulated at www.seer.cancer.gov. © Michael DeGregori

microenvironmental-driven changes that might increase the selective value of the initiating or subsequent oncogenic mutations.

As an example, we can consider how ETV6-RUNX1 translocations could initially expand in the HSC pool by drift, leading to detection of these clones in around one of a hundred newborns—which reflects the chance occurrence of the chromosomal translocation together with clonal expansion via drift. Fortunately, for 99 percent of these newborns, this clone will end up being suppressed, and no leukemia will develop. However, in the 1 percent who do develop ETV6-RUNX1-driven leukemia, delayed exposure to an infection under the conditions described in the previous section may have led to selection for the ETV6-RUNX1 clone, further expanding it and thus facilitating the odds that the next oncogenic mutation occurred. As also described above, certain alleles for immune genes that promote immune hyperresponsiveness, perhaps selected during ancestral times in response to new pathogenic insults (thus representing a disconnect between our ancient genes and modern environments), could modulate the inflammatory response and thus leukemic clonal evolution. Thus, drift-driven expansions can create opportunity for selection to act, which reveals interacting contributions of chance, environment, and inherited genetics in childhood leukemia risk.

In a small way, a better understanding of childhood cancers based on all of the ideas and models described above might provide some comfort to the parents of children who develop these cancers. If any or all of these models are correct, then childhood cancers are not readily preventable. These cancers really may due to be bad luck, at least to a large extent. Thus, these models do not provide easy solutions. There is a possible exception: if the delayed infection model is correct, then earlier exposures to immune stimulants, such as other babies or vaccines, might lessen the numbers of childhood leukemias. As for controlling drift-driven expansions, there is no clear solution, as we cannot change the reality of HSC pool generation—it must start small. Similarly, if childhood cancers are the costs of certain adaptations, like improved brain and immune function, we are stuck with the genetic makeup that we evolved in different times and circumstances.

CHAPTER 15

Evolution-Informed Strategies to Combat Cancer

C ANCER IS A FORMIDABLE opponent. At the time of diagnosis, each cancer can consist of almost a trillion cells that have already evolved the ability to expand and spread in the body, without regard to the health of the host. Unfortunately, such a large population of cells, often with increases in mutation rates, allows the cancer to adapt to new challenges, which facilitates both metastatic spread and the development of therapeutic resistance. While humans evolve defenses slowly, having only several generations per century, cancers can evolve over weeks, benefited by population sizes that dwarf the entire global population of humans. This problem is not unique to cancer, as evolved resistance has hampered our efforts to eliminate the parasites that cause malaria, pathogenic bacteria, mosquitos, crop pests, and so forth. We have our own formidable weapons, including an adaptable immune system and half a billion years of animal evolution to suppress cancer development. For fighting cancers that do develop, we also have a key human trait: our intelligence. We can predict the future, albeit imperfectly, while cancers have no foresight. Cancer has no goals or agenda but simply adapts to current microenvironmental demands. We should be able to steer its course if we could learn how to decrease the odds of resistance development and/or manipulate its microenvironment appropriately to favor less malignant phenotypes. The cancer research community will need to better understand how the microenvironment affects the fitness effects of mutations, a central component of adaptive oncogenesis theory. A major goal should be to identify the key features of the microenvironment that limit tumor genesis in youth but promote oncogenic adaptation when altered with age or insult. Accordingly, cancer biologists will need to move beyond the current mutation-centric paradigm.

The War on Cancer

The National Cancer Act, which initiated what became known as the War on Cancer, was passed in 1971 in the administration of President Richard Nixon. The law provided greater autonomy and substantially more money to the National Cancer Institute in the National Institutes of Health. The annual National Cancer Institute budget subsequently rose from about $0.5 billion in the early 1970s to about $5 billion in 2001, remaining at this level through 2016 (National Cancer Institute 2016). This investment led to significant progress, and our knowledge of the mechanisms underlying cancer development and maintenance has increased exponentially. Progress included the identification of dozens of oncogenes and tumor suppressor genes and the characterization of the intricate cell-signaling and transcription pathways that they control. There have been successes in the clinic, such as the development of imatinib for chronic myeloid leukemia and major improvements in outcomes for children with acute lymphoblastic leukemia. In addition, campaigns to reduce smoking are drastically reducing the rates of smoking-associated cancers, and the early detection of breast, cervical, and colon cancers through screening programs has reduced mortality from these diseases (Figure 15.1). Hormonal and antibody treatments for breast cancers have also improved outcomes.

However, progress for most cancer patients, particularly those diagnosed with cancers at advanced stages, has been unsatisfactory. Outcomes for adults diagnosed with acute myeloid leukemia (AML) are hardly better than they were forty years ago. Even today, most AML patients will die of their disease within two years. For lung cancers, while targeted therapies such as using inhibitors of EGFR have extended survival by around six

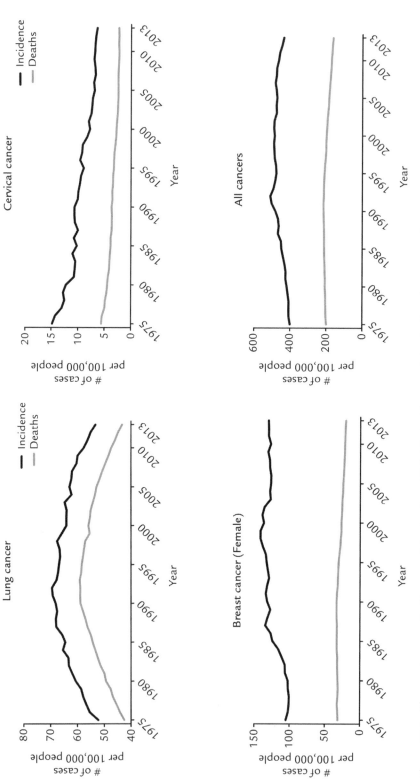

Figure 15.1. Changes in cancer incidence and deaths since 1975. The incidence and deaths from lung, cervical, breast, and all cancers are graphed from data tabulated at (www.seer.cancer.gov). These changes reflect the success of public campaigns to reduce smoking (which affects lung cancer), improvements in early detection (cervical cancer), and improvements in therapy (breast cancer). However, the "All cancers" graph illustrates that overall cancer incidence has not improved substantially, and that while decreasing, death rates from cancer remain high. © Michael DeGregori

months for a fraction of patients with the appropriate driver mutations, and with less of a toll on the patient's quality of life, only a small fraction of patients will survive two or more years following diagnosis. While reductions in smoking have led to large decreases in lung cancer incidence, the diagnosis of lung cancer is still a death sentence for most.

The War on Cancer was initiated during a period of intense excitement in biology as a result of what has been called the molecular revolution. The discovery of the molecular basis of genetics in the 1940s and 1950s led to the development of recombinant DNA technology in the early 1970s. Genes could be transferred between cells and organisms, mutations created, and gene functions elucidated, all of which catalyzed a renaissance in our understanding of cancer cells. However, given the available tools at the time, cancer biology was largely reductionist. The focus was primarily on cell-intrinsic features of cancer—how does oncogene X affect and control cancer cell phenotype Y? Research focused on understanding how oncogenes and tumor suppressor genes affected cancer cells, consistent with the still dominant theory that cancer formation is limited by the occurrence of mutations in these genes. With the understanding that a particular type of cancer is driven by oncogene X, the standard treatment approach is to inhibit the product of oncogene X, following the simple logic that if the cancer is addicted to this oncogene, the cancer cells will die in its absence. As described above, this approach worked beautifully for chronic myeloid leukemia through inhibition of BCR-ABL, and it has improved the length of survival, even if not overall survival, for patients with EGFR-mutant lung cancers. Still, for most cancer patients receiving targeted therapies, responses are limited, and relapse is inevitable.

Given that billions of dollars have been invested, why hasn't more progress been made for cancer patients? We need to keep in mind that the progression from discovery to clinical impact is typically very slow. A target needs to be validated in multiple systems, including animal models, and then a drug needs to be developed. The drug needs not only to inhibit the target but also to have the right properties to do so in a person. For example, the drug cannot be excreted too rapidly or become metabolized into a nonfunctional form before it successfully engages its target in the cancer. The treatment must also have acceptably low toxicity and side effects. Trials to test the potential of the drug in patients can take years to be approved and many more years to complete, at a cost of hundreds of millions of dollars. Through this process, most drugs fail, never being approved for clinical use by the Food and Drug Administration (FDA). In all, the time from the discovery of a target to FDA approval of a therapy can be much more than a decade. Nonetheless, the lack of progress in the search for cancer cures cannot be so easily excused.

A key component of the failure of the War on Cancer to improve outcomes for patients stems from the somatic evolutionary potential of cancer—once it reaches a certain size, a cancer has enormous potential to evolve to evade therapies. We will discuss how large cellular populations of cancers, high mutation rates, and strong selective pressures instituted by therapies engender ideal conditions for somatic evolution of resistance.

An additional, and perhaps very consequential, part of the failure to improve patient outcomes may have resulted from the field's misunderstanding of cancer and the evolutionary parameters that drive its development, progression, and maintenance. The mutation-centric paradigm of cancer has fostered the development of mutation-centric therapies—treatments that target cancer cell phenotypes. As I argued in Chapter 5, this paradigm is inconsistent with available data and evolutionary theory. If we instead understand that the genotypes and phenotypes of cancer have been selected based on microenvironmental demands, then interventions that target the tissue landscape features that underlie the adaptiveness of these mutations and phenotypes would logically be pursued. New theories, such as adaptive oncogenesis, could serve as foundations for these approaches. Strategies to alter the microenvironment in a way that renders particular oncogenic mutations no longer fitness-promoting to the cancer cells should lead to more durable remissions, improving outcomes for patients.

Cancer Therapies Affect Tissue Landscapes

Most conventional chemotherapies target DNA or DNA replication machinery, with the greatest killing of dividing cells, but they also damage tissues with high turnover rates like the hematopoietic system and the intestines. The effectiveness of chemo and radiation therapies are based on the failure of cancer cells to stop and repair. A fast car with no brakes might do fine on the open highway, but it is not likely to survive in a construction zone. Due to reduced checkpoints to halt cell replication following damage, a property intrinsic to many cancers, cancer cells can be particularly sensitive to chemo and radiation therapies. Conventional chemotherapy using decades-old drugs is still the standard of care for most cancers, although it is hoped that the development of more effective and less toxic therapies will eventually relegate these damaging treatments to the history books.

While chemo and radiation therapy can cause the cancer mass to shrink dramatically, the majority of patients with advanced cancers will have only a short remission, from months to a few years. When the cancer comes back, it is typically more aggressive and resistant to treatment. In fact, for some cancers, the therapy buys the patient only a few months of life, which will often be spent suffering the undesirable effects of the therapy. The measure of success of a new therapy is typically judged by how well it reduces cancer burden, with less emphasis on the ability of the therapy to result in long-term suppression of the cancer. Even targeted therapies have not resulted in the hoped-for prolonged remissions for most cancers. Both conventional and targeted therapies, all of which exploit cancer-intrinsic phenotypes, fall within the framework of the mutation-centric paradigm for cancer that been dominant in the past half-century. While these therapies can provide temporary results, they fail to address the underlying environment that favors the malignant phenotype. More frequently than not, the cancer returns.

One problem is that since both conventional and targeted therapies eliminate the majority of the cancer cells, they create strong selective pressure for the acquisition of mutations that provide resistance to the therapy. The mutagenic nature of these therapies further boosts the odds of such resistance. In addition, chemo and radiation therapies raze tissues, causing permanent damage to normal stem cell pools, the supportive cells of the stem cell niche, and tissue cells in general. In addition to causing detrimental side effects, these therapies create severely altered tissue environments that will promote selection for adaptive phenotypes. As discussed above, oncogenic events can be adaptive within such altered environments. Not only can these therapy-altered environments favor the regrowth of residual cancer cells, but they can even promote selection for more aggressive cancer cells that can thrive on the decimated tissue landscape.

Chemo and radiation therapies are also associated with increased risk of secondary cancers, with origins independent of the first cancer. Patients treated with such DNA-damaging therapies, such as for Hodgkin's lymphoma or breast cancer, are at about a fivefold increased risk of developing AML, accounting for 10–20 percent of all AMLs (Morton et al 2013). Therapy-related AML typically develops 3–5 years after the therapy for the first cancer and is almost invariably lethal. While the cancer field has ascribed this increased risk to the mutagenic nature of many chemotherapeutics and radiotherapy (see Godley and Larson 2008), the inhibition or elimination of many normal cells should institute strong selective pressure for cells with mutations that confer resistance to this context. Some of these mutations could be oncogenic. In fact, a recent study indicated that chemotherapy can lead to AML development from preexisting hematopoietic stem cells (HSCs) with mutations in the p53 tumor suppressor gene (Wong et al. 2015). The therapy did not cause the p53 mutations, but it created a context that led to the selection of HSCs with p53 mutations, consistent with previous demonstrations that p53 loss provides HSCs with resistance to radiation-induced apoptosis (Bondar and Medzhitov 2010; Marusyk et al. 2010). In addition, damaged microenvironments, even for tissues like the bone marrow that are distant from the site of the original cancer, can select for oncogenic events within stem cell pools that provide an adaptive advantage, consistent with the theory of adaptive oncogenesis.

Just as strategies that promote a healthy neighborhood can reduce crime, so methods that move the microenvironments that surround cancers back toward greater normalcy—the evolved type—should lead to better control of cancer over years. For a patient with advanced cancer, a treatment regimen may still need to start with methods to eliminate the bulk of the cancer, such as through surgery or drugs. Still, a strategy that considers the impact on the microenvironment will be key.

Therapies targeting cancer cells that reduce the negative impact on noncancer cells need to be developed. Newer immune-based therapies unleash the potential of T-cells to seek out and destroy cancer cells. These tumor-targeting T-cells replicate to generate more of themselves. Immune targeting of cancers is producing amazing results, with long-term remissions for some patients with some metastatic cancers (Khalil et al. 2016). However, these immune therapies can lead to serious autoimmune side effects in some patients, as the same T-cells can sometimes also target normal cells. Still, the recent successes in immunotherapies highlight the potential power of a microenvironment-directed therapy, by changing an immune-suppressive microenvironment into an immune-stimulatory one. From an ecological perspective, these strategies represent a predator-based solution to a pest. If immune responses to cancers can be properly harnessed, increasing cancer targeting while minimizing targeting of noncancer cells, these therapies could offer solutions for patients with currently untreatable cancers.

Looking forward, we need to develop interventions that directly support the microenvironment, with the goal of improving the health of normal tissue cells and structure. Therapies could be designed that protect normal tissue cells from the impact of anticancer therapy, or that promote their regeneration following therapy—a sort of microenvironmental Marshall Plan. Microenvironmental therapies could also be designed that favor more benign cancer stem cells (CSCs) over more malignant ones: even if we cannot achieve a cure, we might be able to tame the cancer toward a phenotype that we can live with. Such interventions should include diet and exercise plans, if such regimens could be shown to promote better maintenance or restoration of the tissue microenvironment during and following therapy. Knowing that we have evolved stem cells to be well adapted to youthful tissue architecture, our goal should be to design interventions that favor the healthy microenvironmental state to which normal cells are best adapted.

Squelching Resistance

The power of selection to lead to the development of resistance to drugs or other human interventions is abundantly clear. Bacterial resistance to antibiotics has become a major public health problem, and the development of each new drug to combat malaria is followed within a few years by the evolution of resistance to the drug by parasites that cause the disease. Cancer is no exception, and the development of resistance to therapies both conventional and targeted often occurs within months or a few years of treatment initiation. For standard cell-toxic chemotherapies, resistance mechanisms can include gene mutations that prevent cell death following damage or the avoidance of the damage by increasing the levels and activity of cellular pumps that expel drugs from the cells. For targeted therapies, drug treatment can mediate selection for cancer cells with mutations that activate other compensatory pathways. Even more commonly, the gene encoding the protein targeted by the drug can be mutated so that the protein no longer binds the drug. In all cases, large population sizes with strong drug-induced selective pressures, often potentiated by higher mutation rates, combine to create a perfect storm for adaptive evolution.

Is resistance inevitable? Not necessarily. We can use the principles of evolutionary biology to design strategies to limit or at least control the development of resistance. Note that these ideas are not unique to adaptive oncogenesis but are logical extensions of a basic understanding of the evolutionary forces of mutation and selection. These strategies include:

Reduce cancer population size. The evolution of therapy resistance requires the presence of cancer cell genotypes that confer such resistance. While mutations conferring resistance can be acquired during therapy, they are often present in the cancer cell population before therapy is initiated. The larger the cancer cell population, the more likely such genotypes will be present. Early detection, when the tumor is small, represents the best opportunity to intervene. Even for large tumors, surgery or other treatments can reduce cancer cell numbers and can be used to reduce the tumor's bulk prior to treatment with the therapy for which resistance development would be the greatest problem. However, debulking tumors using chemotherapy could contribute to the selection of clones resistant to the subsequent therapy, particularly if the resistance method is somewhat general—for example, cellular pump activity that eliminates multiple drugs from cells.

The one-two punch. With populations that can approach a trillion cells, cancers can possess almost every possible resistance mutation even without any increase in mutation rate. For example, if the frequency of a resistance mutation is 10^{-8}, then a cancer with 10^{11} cells will have a thousand such resistant cells. However, if two drugs are used simultaneously that would require completely different resistance mutations, then the chance that the cancer has a cell with both mutations would be the product of the probabilities (10^{-8} and 10^{-8}, 10^{-16}). The presence of such a cell would be highly unlikely even in the largest cancer. The efficacy of such combination therapies has been fantastic for combating AIDS, for which the development of resistance to single-drug therapies was incredibly rapid: the advent of triple-drug combinations is keeping people who are infected with the human immune deficiency virus (HIV) in disease remission for decades. However, even though combination therapies for cancers are used, with more targeted combinations in development, in many cases the same mechanisms mediate resistance to multiple therapies. Thus, while remissions can be longer, cures are still out of reach for most patients with advanced disease. Careful consideration of the evolutionary parameters like cell population size, mutation rate, and mechanisms of resistance for different drugs could allow for the design of more effective therapies.

Target the cancer stem cells. Resistance to therapy can be mediated by selection for cells with mutations conferring such resistance. In addition, as discussed in Chapter 13, CSCs can be intrinsically resistant to both conventional and targeted therapies. Treatments can eliminate the bulk of the cancer, while increasing the fraction of the tumor represented by CSCs. These CSCs can then repopulate the cancer, leading to the patient's relapse. Many current therapies are in essence cutting off the stalk of a weed without removing its roots. Just as a gardener needs to remove the whole weed, or even employ a strategy to selectively kill the root, cancer researchers are trying to develop drugs that better target CSCs (Takebe et al. 2015). One such drug has been approved by the FDA: vismodegib targets the Hedgehog pathway, which is a key pathway in controlling stem cell maintenance. This drug has improved outcomes for patients with advanced basal cell carcinomas, a skin cancer. These cancers have mutations that activate the Hedgehog pathway, leading to pathway addiction and greater sensitivity to drug inhibition. Since the drug targets this addiction, there will be strong selection for resistance. Interestingly, this drug is based on a naturally occurring Hedgehog pathway inhibitor, cyclopamine, which was discovered in the California corn lily. Sheep eating this lily delivered lambs with developmental defects, including possession of a single central eye. These developmental effects limit who can be treated with the drug. Of course, since the Hedgehog and other stem cell pathways are important for the maintenance of normal stem cells even in adults, these drugs must be used judiciously, with the goal of finding dosing or delivery schedules that minimize their impact on normal stem cells while maximizing the impact on CSCs.

The cost of resistance. For bacteria, antibiotic resistance comes with a cost, in that a bacterial clone that acquires antibiotic resistance will typically be outcompeted by other bacteria in the absence of the antibiotic (Melnyk, Wong, and Kassen 2015). However, with time (and generations of bacteria) this cost will usually be lost with the acquisition of additional mutations that compensate for the resistance mutation. While there are fewer data for cancer cells, evidence suggests that at least some resistance mechanisms come with a cost for them, too, which can be observed in the clinic. The T790M mutation in the EGFR gene confers resistance to inhibitors that target this kinase. Kinases are enzymes that relay signals through a cell, such as to proliferate and/or survive, and this mutation prevents a drug from binding to the kinase. Hence, while patients with activated versions of the EGFR gene initially respond to therapy, leading to many months of remission, over half of these patients will relapse with cancers bearing the T790M mutation. Notably, when some patients with such T790-bearing lung cancers are no longer treated with the kinase inhibitor, thus abrogating selection for T790M-mediated resistance, the cancer cells lacking the T790M mutation can return to dominance (Sequist et al. 2011). Thus, the patient can be retreated with the kinase inhibitor originally used. While this example is anecdotal, understanding this cost of resistance should aid in the design of improved therapeutic regimens that capitalize on this cost.

Similarly, mutations that increase the cellular pump activity described above can be very advantageous in the presence of a drug but should be costly in its absence, as these pumps require extra energy in the form of adenosine triphosphate (ATP) to function. Robert Gatenby and colleagues asked whether they could further increase this cost. They provided cancer cells with a nontoxic chemical that uses the pump, so that the drug-resistant cancer cells with high pump activity are burning ATP (depleting their energy stores) at a much higher rate than drug-sensitive cancer cells without the high pump activity (Kam et al. 2015). Since the chemical is nontoxic, ignoring it would be the advantageous strategy, which is what drug-sensitive cancer cells and normal cells with low pump activity inherently do. The nontoxic chemical engenders a cost to the resistance mechanism. Whether this strategy can lead to better maintenance of drug resistance in human cancers remains to be demonstrated.

Carlo Maley and colleagues described such a strategy as a sucker's gambit (Maley, Reid, and Forrest 2004). Their mathematical modeling shows that methods that favor less malignant cells should be useful for prolonging therapeutic efficacy and even eliminating cancers. Other approaches to increase the cost of the resistance phenotype could be developed and would be aided by large-scale screening for drugs that selectively kill resistant rather than sensitive cancer cells. The overall strategy would be to treat the cancer patient with such resistant-cell eliminators that disfavor (and hopefully eradicate) drug-resistant cancer cells, followed by treatment with the original drug to which the cancer should now largely be sensitive. Multiple rounds of such a strategy may be required. These strategies capitalize on our ability to predict resistance mechanisms, which have already been well described for many drugs. In contrast, the cancer cannot predict the future (that is, what we will hit it with next). To be successful, more effort will need to be invested in identifying molecular dependencies that are specific to drug-resistant cancer cells and finding the means to target these vulnerabilities.

Understanding Cancer's Addiction

A number of ice fish species of the family Channichthyidae found in the Southern Ocean around Antarctica are the only known vertebrates that lack hemoglobin and red blood cells (RBCs). In fact, their hemoglobin genes are riddled with mutations that prevent the expression of functional proteins. Given that oxygen solubility is high at cold temperatures, hemoglobin is unnecessary, and lacking RBCs makes for thinner blood that is easier to pump at low temperatures. As described by Sean Carroll in *The Making of the Fittest* (2006), weakened purifying selection on genes that lack adaptive value in the current environment will lead with time to the irreversible degradation of these genes, which is evident in the hemoglobin genes of these fish. In addition, there may have been some adaptive value in losing RBCs in this cold environment, together with positive selection for an antifreeze protein in blood and body fluids. As the earth warms, there is concern that these fish may be doomed, as it will not be possible for them to evolve hemoglobin and RBCs again, given the extensive degradation of the required genes. These fish are permanently adapted for the cold.

Analogously, cancers have been selected for adaptation to the environments that shaped their evolution, without contingency plans for possible changes to this environment. As explained by Fisher's geometric model (1930), discussed in Chapter 8, adaptation to a new environment necessarily entails reduced adaptation to the original environment. As an extension of the theory of adaptive oncogenesis, we can hypothesize that restoration of a more normal microenvironment, akin to relocating ice fish to warmer waters, would result in cancer cells that were poorly adapted to the restored microenvironment and could even stimulate evolution of the cancer toward a more benign phenotype (one better in tune with the normal environment).

Cancers also acquire oncogenic mutations that lead to hyperactive enzymes, like the BCR-ABL kinase discussed in Chapter 5. Such mutations create addictions, as BCR-ABL mutated leukemias require the activity of this kinase to a much greater extent than noncancer cells require the normal ABL kinase. The inhibitor of this kinase, imatinib, has been hugely successful clinically, exploiting the strong addiction of these leukemias to BCR-ABL activity. While most patients with these leukemias can be treated with imatinib indefinitely, keeping their leukemia cells at very low numbers, resistance can develop in a fraction of patients. Resistance development typically involves mutation of the BCR-ABL gene so that imatinib no longer binds and inhibits the BCR-ABL kinase (Gorre and Sawyers 2002). Fortunately, newer drugs have been developed that can inhibit these resistant forms of BCR-ABL. However, the ability to control BCR-ABL-driven leukemias with a single drug is an exception. For most cancers, including other leukemias, even effective inhibition of the oncogene to which the cancer is addicted produces only transient remission. Cancers are very effective at evolving resistance to therapies, whether conventional or targeted. In the next section, I discuss methods that reduce the development of resistance by altering fitness landscapes.

Tissue Landscape Management

A full comprehension of the forces that determine cancer risk, shape cancer development, and underlie its responses to therapies is essential for developing strategies to diagnose, prevent, and cure cancer. As emphasized in the theory of adaptive oncogenesis, strategies need to consider how tissue microenvironments dictate the fitness value of precancer and cancer genotypes.

Given the current mutation-centric paradigm, molecular methods for the early detection of cancers have primarily focused on detecting mutations. If cancer development is less limited by mutation occurrence than it is by alterations in the microenvironment, tests should also assess the quality of the tissue landscape as a measure of cancer risk. To this end, studies need to identify the parameters of the landscape that best predict the propensity for somatic evolution, which should be a critical step toward improved early diagnosis. Elucidating these parameters could also lead to screening approaches to identify interventions, whether dietary or pharmaceutical, that push critical parameters back toward their evolved values.

Similarly, while current methods employed to identify a chemical as a carcinogen rely on the ability of the chemical to cause mutations, we should ask whether the ability of a particular exposure to perturb tissue landscapes could provide improved or complementary methods for identifying carcinogens. As an example, regular consumption of very hot liquids (such as yerba maté tea of South America) is associated with significantly increased risk of esophageal cancer. Studies have shown that the temperature of the water, not the extracted tea or coffee, is the culprit (Islami et al. 2009). According to adaptive oncogenesis, thermal damage to the throat, with associated inflammation and cycles of tissue repair, should substantially alter the adaptive landscape for oncogenic mutations. More generally, we can ask whether the propensity to impair tissue health can be used to identify suspected carcinogens, as well as to understand their mechanisms of action and thus develop mitigating interventions.

Most current therapies target cancer cells, with the goal of eliminating as much of the cancer as possible with the least negative impact on the patient, but they can still take the patient to or even beyond the brink of death. This approach is followed in the development of new anticancer therapies and their dosing regimen. Clinical trials for new drugs proceed through three phases, with phase I determining tolerated dosing, phase II determining efficacy relative to previous experience with standard-of-care drugs, and phase III performing a direct comparison with the current therapy. Primary end points for phase II and III clinical trials are objective tumor responses—basically, tumor shrinkage.

Currently, phase I clinical trials for a new anticancer drug are designed to determine the maximal tolerated dose (MTD), which in the United States is the dose where less than one-third of patients experience excessive drug-induced toxicity (Le Tourneau, Lee, and Siu 2009). The MTD is typically the recommended dose for a phase II trial, as well as the dose often employed for approved drugs in the clinic. The assumption is that efficacy and toxicity both increase with dose. For cells in a dish, this assumption is likely true. However, for cancer cells in a person, using the MTD ignores the ecology of cancer. The impact of the treatment on the microenvironmental landscape for the cancer is disregarded. Not only does using the MTD subject the patient to debilitating side effects, including damage to vital organs and immune suppression, but this approach will inevitably select for drug resistance and an aggressive cancer phenotype adaptive to the damaged tissue landscape.

Understanding that the fitness values of the cancer phenotypes engendered by the underlying oncogenic mutations are dependent on the microenvironment should lead to new strategies to treat cancer by altering the tissue states. Ideally, a treatment strategy informed by tissue ecology would consider the complexities of a tissue to best tune the system toward the evolved phenotype. Tuning a microenvironment will undoubtedly require subtler manipulations than trying to kill the cancer. In fact, some of this tuning may be to correct tissue damage resulting from the therapy directed against the cancer.

We can learn from agriculture. Integrated pest management is an agricultural strategy to control, but not eliminate, pests that damage crops, thus reducing the amount of pesticide used. Pesticide is applied as needed to minimize the economic impact of the pests, with applications based on monitoring of the pest (whether insect, fungus, or weed) populations in the fields. Only when pest levels exceed some threshold are control measures applied. As discussed above, the development of resistance has a cost, since the phenotypic change required for resistance entails movement away from the previous phenotypic optimum (Figure 8.1). Under intense treatment, the benefit of the resistance phenotype will outweigh the costs. However, at lower levels of treatment or with intermittent application, the development of resistance can be limited or even prevented, as costs will balance benefits. Nonresistant genotypes will persist and even rebound during periods between treatments, preventing dominance by the resistant genotypes. Careful maintenance of healthy crops is another key feature of integrated pest management. Finally, early and excessive application of pesticides leads to greater killing of the natural predators of insect pests, such as birds, frogs, wasps, and spiders, thus removing natural control mechanisms.

Gatenby and colleagues have leveraged a very similar strategy for the control of cancer. Using mathematical modeling, they developed treatment schedules that they predicted would maintain chemosensitive cancer cells, given their fitness advantage over chemoresistant cancer cells (at least in the absence of treatment). Their method, called adaptive therapy, involves the treatment of the patient only to the point where the tumor burden is reduced to a nondisruptive level (that is, one tolerated by the patient) (Gatenby et al. 2009). The regimen is adaptive, as cancer burden is monitored in the patient and therapy applied only as needed to control tumor burden, not to eliminate it. It is integrated pest management, but the pest is cancer. Strikingly, this approach works in mouse models of breast cancer, with reductions in chemotherapy dosing over time (Enriquez-Navas et al. 2016). Reduced intensity of treatment will not only decrease selection for resistant clones but will also better favor the normal tissue landscape (a healthy crop). In addition, just as integrated pest management can reduce the killing of natural predators of pests, reduced chemo application should better spare immune cells, which can serve as predators for cancer cells. Adaptive therapy is now in clinical trials for prostate cancer. Solid evidence of efficacy in humans will be required to advance this approach. Perhaps just as important, acceptance of this approach will require abandoning the "more is better" philosophy in medicine, whether in the sense of more treatment (using the MTD) or more initial elimination of the cancer burden (objective responses).

New therapeutic approaches need to leverage a key concept: cancer has no purpose or direction. Cancer cells are simply adapting to current microenvironmental conditions. By manipulating the microenvironment, we should be able to direct cancer evolution without directly attacking the cancer cells. Indeed, studies in mice have shown how normalizing oxygen or pH can reduce the aggressiveness of a cancer, improving outcomes (Mazzone et al. 2009; Ibrahim-Hashim et al. 2017). These studies serve as proof of principles for the concept that altering the microenvironment, even a single component like oxygen or pH, can direct cancer evolution toward more benign phenotypes.

Cancer relapse following therapy, particularly the more toxic conventional therapies, typically manifests itself in a form that is more aggressive and harder to treat. As discussed above in this chapter, therapy-mediated damage to the microenvironment promotes selection for cancer phenotypes adaptive to this environment, and phenotypes adaptive to damaged landscapes tend to be more invasive and insult-resistant. Recent studies by Judith Campisi and colleagues have provided support for this idea, and even a potential Marshall Plan to restore landscapes after chemotherapy. Chapter 12 introduced mouse models that allow the deletion of senescent cells, using either transgenic approaches or drugs. Campisi and collaborators used these models to show that chemotherapeutics induce cellular senescence, contributing to systemic inflammation (as previously shown for aging) (Demaria et al. 2016). Importantly, clearance of senescent cells reversed multiple negative effects of chemotherapy in mice, including hematopoietic suppression, cardiac dysfunction, and reduced physical activity or strength. Strikingly, the deletion of senescent cells with consequent reductions in inflammation also reduced cancer recurrence and metastases following chemotherapy: restoration of a more normal (pretherapy) landscape disfavored an aggressive cancer phenotype! Of note, chemotherapy-induced senescent cells that secrete inflammatory cytokines are also observed in breast cancer patients, suggesting that such an approach might be useful in humans (Sanoff et al. 2014).

Basic research into the salient features of perturbed tissue landscapes that favor oncogenesis will improve both cancer prevention and treatment strategies. For example, as we discover differences between young and old tissues and learn how these differences promote somatic evolution in the old tissues, we will be able to ask whether reversing these differences (toward the more youthful phenotype) can limit cancer development, as shown through targeting inflammation (Henry et al. 2015). A better appreciation of the role of tissue landscapes in cancer development and maintenance should lead to strategies aimed at targeting tumor cells with a reduced impact on normal tissue architecture and function, and to the development of interventions that directly favor the more normal phenotypes. These strategies will require research to elucidate the microenvironmental factors that dictate the fitness effects of particular oncogenic mutations, followed by screens to identify interventions that can switch these parameters from fitness-promoting to fitness-reducing for these driver mutations.

As described in Chapter 9, lung cancers can evolve through the acquisition of mutations in multiple different kinases, including EGFR, and other signaling proteins (Hirsch et al. 2016). I have posited that such mutations are selected due to changes in the lung landscape, whether resulting from aging or smoking. These cancers are addicted to these mutated genes, as the mutations were instrumental in their reaching an adapted state. We can envision two general, and not mutually exclusive, strategies to exploit evolved oncogenic addictions: target the oncogene and target the microenvironmental change that led to oncogenic adaptation.

For the first strategy, drugs targeting EGFR, like erlotinib, have increased the survival of patients with EGFR-mutated lung cancers relative to conventional chemotherapeutic treatments (Herbst and Bunn 2003). In addition, patients better tolerate these kinase inhibitors, which leads to an improved quality of life. Similar successes have been achieved for patients with lung cancers bearing other mutated kinase genes.

However, almost all patients will eventually relapse, typically within 2–3 years of diagnosis, and will die from their disease (West, Oxnard, and Doebele 2013). As discussed above, for about half of EGFR-mutant cancers, the relapse bears a mutation in the EGFR gene (T790M) that prevents erlotinib from binding. A drug recently approved by the FDA that can also inhibit this resistant form of EGFR leads to longer remissions, although the cancers still mutationally exploit other pathways to develop resistance. Given the large numbers of cells and high mutation rates, lung cancers are potent reservoirs of genetic diversity upon which treatment-mediated selection can act. Drug-mediated targeting of the oncogene will inevitably create strong selective pressure for new adaptive mechanisms.

For the second strategy (informed by adaptive oncogenesis), we can consider that the advantage conferred by EGFR or other driver mutations is dependent upon the microenvironmental context that promoted adaptation by these oncogenic events. The goal would be to manipulate the microenvironment so that mutated EGFR is no longer adaptive. If this therapeutically restored microenvironment was more youthful-like, other oncogenic events should be similarly nonadaptive. Therefore, there will be selection for less malignant phenotypes. Such strategies will be particularly relevant for treating cancers driven by oncogenes, such as mutated KRAS, that are not inhibited by currently available drugs. Key to this strategy will be to understand which tissue landscape changes due to age or carcinogenic insult promote selection for the oncogene, and how we can reverse these changes to mitigate such oncogenic adaptation. It will be easier to achieve the former goal than the latter. For example, for KRAS in lung cancer, we would need to determine the features of the smoked lung landscape that promoted selection for KRAS mutation. We would then need to uncover interventions to change these features so that KRAS-activating mutations were no longer adaptive. Fully restoring tissue landscapes to undamaged or more youthful phenotypes will be difficult, although we may discover modulatable parameters (such as particular inflammatory cytokines, pH, or senescent cells) that can change the selective value of cancer genotypes. Perhaps reversing one or a few of the microenvironmental factors, even without reversing many others, could affect the fitness value of oncogenic mutations to the extent that cancer patients would have durable remissions.

As an additional microenvironmental strategy, immune-based approaches can also decrease the fitness of cancer cells by facilitating their selective elimination by T-cells (Khalil et al. 2016). Therapies that boost the ability of the immune system to target cancers are showing striking efficacy in a subset of patients with lung cancers. In this case, the high mutation burden of lung cancers is a good thing, as immune cells can recognize these cancer cells as foreign and target them. The intelligent combination of targeted, microenvironment-normalizing, and immune-based strategies may be critical for overcoming the powerful ability of cancers to evolve in response to drug treatments.

Tissue ecology landscaping should be less damaging to the patient, as successful interventions will likely disfavor cancer phenotypes by moving the tissue landscape toward greater normalcy. This approach could also better address tumor heterogeneity, as normalization of the microenvironment could reduce the fitness of various clones bearing distinct oncogenic mutations that provide adaptations to different tissue perturbations. For example, could lessening inflammation or normalizing tumor vasculature improve the status of multiple cancer neighborhoods? Treating metastatic cancers will create additional challenges, as there will not be one normal state to favor, since multiple tissues will be plagued by the cancer. Ecological strategies will not be easy, as currently employed large-scale screens for compounds that kill cancer cells may be more feasible than screens for landscape-normalizing factors. Nevertheless, since we have almost entirely invested in the former strategy, we do not know how difficult the latter will be.

Understanding how tissue and tumor microenvironments affect the trajectories of malignant somatic evolution, and how carcinogenic contexts and therapies change these microenvironments, could lead to new paradigms that connect cancer to its causes and could catalyze the development of interventions to minimize, delay, or mitigate detrimental microenvironmental changes and thus reduce cancer risk, progression, and/or aggressiveness.

CHAPTER 16

A New Framework for Understanding and Controlling Cancer

THEODOSIUS DOBZHANSKY FAMOUSLY SAID "Nothing in biology makes sense except in the light of evolution" (1973). We can safely include all of medical biology in this dictum. Yet the focus of medical education, research, and care has been on proximate mechanisms, not evolutionary (ultimate) causations. Most medical researchers, doctors, and students are unlikely to be aware of the evolutionary reason for why we age and, consequently, why we develop diseases of aging. Similarly, the evolutionary reasons for why we develop cancer are poorly appreciated. Even those who are aware of such connections are unlikely to incorporate these ideas into practice. We do need to understand proximate mechanisms, such as the mutations and tissue changes that can lead to cancer, or the dietary habits that lead to coronary obstruction and heart disease. While appreciating such proximate cause-and-effect relationships can lead to interventions that can reduce the effects of disease, understanding evolutionary causation can provide an important framework that should have an even greater impact. Throughout this book, I have argued that linking cancer incidence with its causes—including aging, exposures, and genetics—requires understanding how these causes influence tissue microenvironments and the fitness impacts of mutations.

An ecological and evolutionary perspective on cancer will be required for the development of improved strategies for prevention and treatment.

Medicine through the Lens of Evolutionary Thought

I have discussed how the intense focus on mutations as the causes of cancer, without considering how tissues changes with age or exposures dictate the fitness value of mutations, has engendered mutation- and phenotype-directed therapies that have not been nearly as successful as hoped. The theory of adaptive oncogenesis emphasizes the role of microenvironmental change resulting from old age or other carcinogenic contexts in engendering greater cancer risk. Recognizing how natural selection has tamed somatic evolution through tissue maintenance, and the limits of this taming that occur, particularly as we age, should lead to strategies guided by evolutionary theory that in turn should produce improved cancer prevention and more durable therapeutic responses. Using evolutionary theory, we can also better understand how modern conditions and lifestyle choices affect somatic tissue systems, for which the underlying genetics evolved under many millennia of very different environmental conditions. Given that disruption of evolved tissue maintenance strategies, whether through smoking, excessive dietary intake, or living longer, can promote disease, we should be able to use this information to better prevent and treat these diseases. Recognizing that natural selection has acted to maximize reproductive success, with consequent trade-offs in terms of the long-term maintenance of our health, is critical for designing interventions to reduce negative attributes without overly inhibiting the positive ones that underlie the evolution of traits like inflammation. An evolutionary understanding of health and disease provides the theoretical structure into which proximate explanations must be incorporated.

Despite the importance of evolutionary biology as the foundation for understanding physiology and disease, educational programs for future biomedical researchers and doctors largely ignore how evolution has determined disease susceptibility and our defenses against it. In medical and graduate schools, there is intense competition from different subspecialties for contact hours with students. Still, the better integration of evolutionary thinking into all areas of biology and medicine should not require more class time, just a restructuring of how physiology and diseases are taught. We need to teach future scientists and doctors to understand everything through the lens of evolution, even if it means not having them memorize a multitude of facts. As Randolph Nesse and George Williams argued in *Evolution and Healing*, incorporating evolutionary biology into medical school curriculums "will give students not only a new perspective on disease but also an integrating framework on which to hang a million otherwise arbitrary facts" (Nesse and Williams 1996, 245). Our education systems need to stress the importance of a comprehensive education in evolutionary and ecological theory at all levels, beginning in grade school. Evolutionary biology provides a framework to understand the ultimate answers to why questions.

Evolutionary Cancer Theory and the Identification of the Causes of Cancer

An evolutionary and ecological understanding of cancer development will allow scientists and policy makers to better connect cancer incidence with its causes. While we do not need a new theory to know that smoking causes cancers, we need to develop a new theory to explain why this is the case. I have argued that understanding such connections simply as mediated by altering mutation occurrence is incorrect and will prevent us from developing improved interventions. The causes of cancer, whether old age, smoking, or radiation exposure, can alter mutation frequency, which contributes to somatic evolution and consequently cancer risk. However, only by realizing that the fitness effects of these mutations are highly dependent on their microenvironmental context can we truly understand the origins and vulnerabilities of cancer.

Evolution has worked with relatively simple rules that govern mutation, selection, and drift, leading to complex outcomes. Mutations created at each generation engender random phenotypic variation upon which selection and drift can act. Yet from these simple rules have emerged extremophile bacteria, orchids, manta rays, HIV, and humans. In addition to generating the myriad of forms and functions of life on earth, natural selection has acted to limit cancer incidence during years when reproduction is likely, with reduced tumor suppression late in life when the influence of natural selection wanes. Adaptive oncogenesis theory explains how the timing of these changes in selective pressures has been dictated by evolved life histories—the strategies for reproduction, tissue maintenance, and aging that different animal species have evolved, in good measure dependent on external pressures. Somatic cell fitness is a critical part of the overall strategy of an individual to maximize its own fitness, improving the odds of successfully passing its genes on to future generations. Investment in the maintenance of tissues, which allows resident stem cells to remain well adapted and thus fit, is critical not only for delaying physiological aging until periods where reproductive success is less likely, but also for limiting the cancers that can reduce the fitness of the individual animal. Thus, the rates of aging and cancer risk are inextricably linked, scaled to match evolved life histories.

While carcinogenesis is only one form of somatic evolution, its impact on our lives makes our understanding of this process critical. Carcinogenesis follows similar evolutionary principles known from organismal biology. Cell divisions, endogenous damage, and external exposures create mutations and epigenetic changes—the heritable phenotypic diversity upon which selection can act. I have described above how this selection is predominantly stabilizing during youth but transitions to give positive selection for particular mutations a greater role later in life, with the timing dictated by evolved life histories. Life histories determine tissue maintenance strategies that dictate the microenvironments experienced by stem cells, which in turn affects the fitness effects of mutations (Rozhok and DeGregori 2016).

While this new understanding that links cancer with its causes may appear more complex than the simple paradigms of aging → mutations → cancer or carcinogen → mutations → cancer, it is much more consistent with our understanding of the fundamental principles that guide organismal evolution. It is to be hoped that our understanding of the factors that dictate the fate of somatic mutations in animal tissues will help merge cancer and evolutionary theories to create a model of cancer that can better tackle the many unexplained phenomena governing the genesis, development, and therapeutic responses of cancers. This new model should also address how somatic evolution has interacted—and will continue to do so—with evolution at the organismal level.

Limiting the Impact of Cancer

Do we need to understand how cancer evolves to diagnose, prevent, and treat it? The answer is an unequivocal yes. For example, consider how we use experimental systems to learn about cancer formation. There are many model systems, from computational to animal, used to study cancers that are well known to be associated with either aging, smoking, or other factors. However, genetically engineered mouse models to study these cancers rarely use older animals or incorporate exposure to smoking or the other associated contexts. Based on the simple paradigm of aging or carcinogen → mutation → cancer, if the model introduces the associated oncogenic mutations, then the causative context (like aging or smoking) becomes superfluous as a model component. Similarly, most computational models consider only the time component of aging, not the associated physiological changes. Environment matters. Just as it is futile to try to reduce drug addiction in the inner city without considering the underlying social and economic conditions there, so models to understand cancer need to incorporate the contexts underlying the disease and their impacts on tissues. Otherwise, these models will provide insights into cancer that are incomplete or incorrect, and prevention and treatment interventions derived from these models could be misdirected.

Beyond simply targeting cancer phenotypes, we need to learn how to manipulate the fitness value of oncogenic genotypes by controlling the tissue microenvironment. While there can be value in targeting the oncogenic proteins themselves, with few exceptions this approach has not led to durable remissions for patients with cancer. Moreover, conventional therapies kill tumor cells but also decimate host cells and tissues, pushing microenvironments even farther from the evolved phenotype. We need to develop interventions that target not only the intrinsic workings of the cancer cells but also the perturbed microenvironment, which not only promoted the genesis of the cancer but still favors the cancer phenotype.

The National Cancer Institute has undertaken a massive project to sequence thousands of cancers, known at The Cancer Genome Atlas, at a cost of more than $200 million. This project is founded on the mutation-centric paradigm for cancer: understanding cancers and their vulnerabilities by identifying their mutations. While characterizing mutations in cancers has value, such an investment without a parallel effort to characterize the tissue changes that led to the selection for these mutations is likely to prove shortsighted. Another big question is how to distinguish driver mutations in oncogenes and tumor suppressor genes from passenger mutations, out of the thousands of identified mutations in each cancer. Tests for potential driver activity of identified mutations should incorporate the context associated with the particular cancer. Currently, efforts to answer this question occur largely in young unperturbed cells and animals, based on the assumption that oncogenic mutations are largely context-independent drivers of malignancy. However, if oncogenesis is enabled by microenvironmental changes that promote selection for adaptive oncogenic mutations, then the exposures or context associated with the cancer needs to be incorporated.

Determining oncogenic potential by the ability of a mutated gene to induce clonal expansions under the relevant context (adaptiveness) may represent a more accurate reading of the oncogenic potential of a suspected driver gene. The carcinogenicity of a chemical could similarly be investigated by asking whether exposure of animals to the suspected chemical could provide a context that promoted oncogenic adaptation. These short-term experiments would be much more economical and humane than long-term experiments with tumor development in animals. Moreover, if cancer is not limited by the occurrence of oncogenic mutations, then methods to detect early lesions need to focus more on monitoring somatic evolution (clonal expansions) and other markers of a perturbed adaptive landscape, compared to the "present or absent" detection methods for oncogenes.

An evolutionary understanding of cancer, guided by adaptive oncogenesis and other theories, should determine how we prevent, diagnose, and treat cancers. From the mutation-centric perspective, avoiding cancer means limiting exposure to agents that cause mutations. While I would agree that avoiding mutagens should be part of any personal strategy and public policy, an ecological or evolutionary perspective would expand this advice to include avoidance of exposures or contexts that perturb tissues in ways that encourage oncogenesis, and leveraging strategies that best maintain tissues in their evolved states. Maintaining our tissues in more youthful states or returning the landscape of a cancer-affected tissue to a more normal state, to disfavor oncogenic mutations, is easier said than done. However, given the dominance of the current mutation-centric paradigm, there has been little effort to even attempt to achieve these goals.

To manipulate tissue fitness landscapes, we need to better understand how the condition of tissue landscapes can either hinder or promote oncogenesis. Methods need to be developed that can assess the ability of compounds, both synthetic and natural products, to contribute to tissue maintenance. Perhaps more important, we can ask which lifestyles, exercise regimens, and diets best promote the evolved tissue landscape, which will require the development of economical and noninvasive tests for tissue fitness. A solid theory anchored in evolutionary biology that can explain connections between cancer and its causes is needed to frame the questions, experimental methods, clinical interventions, and public policies to best limit the devastating impact of cancer on our lives.

We can marvel at the power of natural selection, which can endow a single fertilized egg with the ability to generate a body of more than a trillion cells and keep that body relatively healthy for decades—all for the purpose of making more bodies. At the same time, to tackle challenges like cancer we must also understand the imperfections and limitations inherent in the human body. Understanding how evolution got things right, such as programmed mechanisms for tissue maintenance, will be important for preventing disease and cancer. At the somatic level, appreciating how cancer evolves within us, and the forces that sculpt this evolution, will likewise be critical, such as for the development of novel therapeutic approaches and for preempting the development of drug resistance in cancers. Recognition that the fitness effects of oncogenic mutations are highly dependent on the tissue microenvironment can allow us to design strategies to disfavor precancer and cancer cells.

Cancer risk has been shaped by evolution at the organismal level, and cancers evolve within us, molded by many of the same evolutionary forces. A deep understanding of both of these frameworks will allow us to better control this dreaded disease.

References

Abegglen, L. M., A. F. Caulin, A. Chan, K. Lee, R. Robinson, M. S. Campbell, W. K. Kiso, et al. 2015. "Potential Mechanisms for Cancer Resistance in Elephants and Comparative Cellular Response to DNA Damage in Humans." *JAMA* 314 (17): 1850-60.

Abkowitz, J. L., S. N. Catlin, M. T. McCallie, and P. Guttorp. 2002. "Evidence That the Number of Hematopoietic Stem Cells Per Animal Is Conserved in Mammals." *Blood* 100 (7): 2665-67.

Adler, M. I., and R. Bonduriansky. 2014. "Why Do the Well-Fed Appear to Die Young?" *BioEssays* 36 (5): 439-50.

Alberts, S. C., E. A. Archie, L. R. Gesquiere, J. Altmann, J. W. Vaupel, and K. Christensen. 2014. "The Male-Female Health-Survival Paradox: A Comparative Perspective on Sex Differences in Aging and Mortality." In *Sociality, Hierarchy, Health: Comparative Biodemography: A Collection of Papers*, edited by M. Weinstein and M. A. Lane, 339-64. Washington, DC: National Academies Press.

Alexandrov, L. B., Y. S. Ju, K. Haase, P. Van Loo, I. Martincorena, S. Nik-Zainal, Y. Totoki, et al. 2016. "Mutational Signatures Associated with Tobacco Smoking in Human Cancer." *Science* 354 (6312): 618-22.

American Cancer Society. n.d. Accessed September 2, 2017. https://www.cancer.org/cancer/cancer-causes/tobacco-and-cancer/health-risks-of-smoking-tobacco.html.

Anand, P., A. B. Kunnumakara, C. Sundaram, K. B. Harikumar, S. T. Tharakan, O. S. Lai, B. Sung, and B. B. Aggarwal. 2008. "Cancer Is a Preventable Disease That Requires Major Lifestyle Changes." *Pharmaceutical Research* 25 (9): 2097–116.

Anderson, A. R., A. M. Weaver, P. T. Cummings, and V. Quaranta. 2006. "Tumor Morphology and Phenotypic Evolution Driven by Selective Pressure from the Microenvironment." *Cell* 127 (5): 905–15.

Anderson, K., C. Lutz, F. W. van Delft, C. M. Bateman, Y. Guo, S. M. Colman, H. Kempski, et al. 2011. "Genetic Variegation of Clonal Architecture and Propagating Cells in Leukaemia." *Nature* 469 (7330): 356–61.

Armitage, P., and R. Doll. 1954. "The Age Distribution of Cancer and a Multi-Stage Theory of Carcinogenesis." *British Journal of Cancer* 8 (1): 1–12.

Arnal, A., B. Ujvari, B. Crespi, R. A. Gatenby, T. Tissot, M. Vittecoq, P. W. Ewald, et al. 2015. "Evolutionary Perspective of Cancer: Myth, Metaphors, and Reality." *Evolutionary Applications* 8 (6): 541–44.

Association of Road Racing Statisticians, World Single Age Records- Marathon. http://www.arrs.net/SA_Mara.htm.

Austad, S. N. 1993. "Retarded Senescence in an Insular Population of Virginia Opossums (*Didelphis virginiana*)." *Journal of Zoology* 229 (4): 695–708.

Bagby, G. C., and A. G. Fleischman. 2011. "The Stem Cell Fitness Landscape and Pathways of Molecular Leukemogenesis." *Frontiers in Bioscience* 3: 487–500.

Bagby, G. C., and G. Meyers. 2007. "Bone Marrow Failure as a Risk Factor for Clonal Evolution: Prospects for Leukemia Prevention." *Hematology* 2007: 40–46.

Baker, D. J., B. G. Childs, M. Durik, M. E. Wijers, C. J. Sieben, J. Zhong, R. Saltness, et al. 2016. "Naturally Occurring p16Ink4a-Positive Cells Shorten Healthy Lifespan." *Nature* 530 (7589): 184–89.

Baker, N. E. 2011. "Cell Competition." *Current Biology* 21 (1): R11–15.

Ballesteros-Arias, L., V. Saavedra, and G. Morata. 2014. "Cell Competition May Function Either as Tumour-Suppressing or as Tumour-Stimulating Factor in Drosophila." *Oncogene* 33 (35): 4377–84.

Barcellos-Hoff, M. H., and S. A. Ravani. 2000. "Irradiated Mammary Gland Stroma Promotes the Expression of Tumorigenic Potential by Unirradiated Epithelial Cells." *Cancer Research* 60 (5): 1254–60.

Beerman, I., W. J. Maloney, I. L. Weissmann, D. J. Rossi. 2010. "Stem Cells and the Aging Hematopoietic System." *Current Opinion in Immunology* 22 (4): 500–506.

Behjati, S., M. Huch, R. van Boxtel, W. Karthaus, D. C. Wedge, A. U. Tamuri, I. Martincorena, et al. 2014. "Genome Sequencing of Normal Cells Reveals Developmental Lineages and Mutational Processes." *Nature* 513 (7518): 422–25.

Berry, R. J., and F. H. Bronson. 1992. "Life History and Bioeconomy of the House Mouse." *Biological Reviews of the Cambridge Philosophical Society* 67 (4): 519–50.

Bissell, M. J., and W. C. Hines. 2011. "Why Don't We Get More Cancer? A Proposed Role of the Microenvironment in Restraining Cancer Progression." *Nature Medicine* 17 (3): 320–29.

Bize, P., F. Criscuolo, N. B. Metcalfe, L. Nasir, and P. Monaghan. 2009. "Telomere Dynamics Rather Than Age Predict Life Expectancy in the Wild." *Proceedings of the Royal Society B: Biological Sciences* 276 (1662): 1679–83.

Blagosklonny, M. V. 2013. "Rapamycin Extends Life- and Health Span Because It Slows Aging." *Aging* 5 (8): 592–98.

Blanpain, C., M. Mohrin, P. A. Sotiropoulou, and E. Passegué. 2011. "DNA-Damage Response in Tissue-Specific and Cancer Stem Cells." *Cell Stem Cell* 8 (1): 16–29.

Blasco, M. A. 2005. "Telomeres and Human Disease: Ageing, Cancer and Beyond." *Nature Reviews Genetics* 6 (8): 611–22.

———. 2007. "Telomere Length, Stem Cells and Aging." *Nature Chemical Biology* 3 (10): 640–49.

Blokzijl, F., J. de Ligt, M. Jager, V. Sasselli, S. Roerink, N. Sasaki, M. Huch, et al. 2016. "Tissue-Specific Mutation Accumulation in Human Adult Stem Cells During Life." *Nature* 538 (7624): 260–64.

Boland, C. R., and A. Goel. 2010. "Microsatellite Instability in Colorectal Cancer." *Gastroenterology* 138 (6): 2073–87.

Bolveri, T. 1929. *The Origin of Malignant Tumors*. Baltimore, MD: Williams and Wilkins.

Bondar, T., and R. Medzhitov. 2010. "p53-Mediated Hematopoietic Stem and Progenitor Cell Competition." *Cell Stem Cell* 6 (4): 309–22.

Bongaarts, J. 2009. "Human Population Growth and the Demographic Transition." *Philosophical Transactions of the Royal Society of London: Series B, Biological Sciences* 364 (1532): 2985–90.

Boshoff, C., and R. Weiss. 2002. "AIDS-Related Malignancies." *Nature Reviews Cancer* 2 (5): 373–82.

Bowie, M. B., K. D. McKnight, D. G. Kent, L. McCaffrey, P. A. Hoodless, and C. J. Eaves. 2006. "Hematopoietic Stem Cells Proliferate Until after Birth and Show a Reversible Phase-Specific Engraftment Defect." *Journal of Clinical Investigation* 116 (10): 2808–16.

Bozic, I., T. Antal, H. Ohtsuki, H. Carter, D. Kim, S. Chen, R. Karchin, K. W. Kinzler, B. Vogelstein, and M. A. Nowak. 2010. "Accumulation of Driver and Passenger Mutations during Tumor Progression." *Proceedings of the National Academy of Sciences of the United States of America* 107 (43): 18545–50.

Bozinovski, S., R. Vlahos, D. Anthony, J. McQualter, G. Anderson, L. Irving, and D. Steinfort. 2016. "COPD and Squamous Cell Lung Cancer: Aberrant Inflammation and Immunity Is the Common Link." *British Journal of Pharmacology* 173 (4): 635–48.

Brandhorst, S., I. Y. Choi, M. Wei, C. W. Cheng, S. Sedrakyan, G. Navarrete, L. Dubeau, et al. 2015. "A Periodic Diet That Mimics Fasting Promotes Multi-System Regeneration, Enhanced Cognitive Performance, and Healthspan." *Cell Metabolism* 22 (1): 86–99.

Brannon, A. R., E. Vakiani, B. E. Sylvester, S. N. Scott, G. McDermott, R. H. Shah, K. Kania, et al. 2014. "Comparative Sequencing Analysis Reveals High Genomic Concordance between Matched Primary and Metastatic Colorectal Cancer Lesions." *Genome Biology* 15 (8): 454. Accessed September 3, 2017. https://genomebiology.biomedcentral.com/track/pdf/10.1186/s13059-014-0454-7?site=genomebiology.biomedcentral.com.

Brown, J. S., and C. A. Aktipis. 2015. "Inclusive Fitness Effects Can Select for Cancer Suppression into Old Age." *Philosophical Transactions of the Royal Society B: Biological Sciences* 370 (1673): 2015160. doi: 1098/rstb.2015.0160.

Buffenstein, R., and J. U. Jarvis. 2002. "The Naked Mole Rat—A New Record for the Oldest Living Rodent." *Science of Aging Knowledge Environment* 2002, no. 21: pe7.

Busque, L., J. P. Patel, M. E. Figueroa, A. Vasanthakumar, S. Provost, Z. Hamilou, L. Mollica, et al. 2012. "Recurrent Somatic *TET2* Mutations in Normal Elderly Individuals with Clonal Hematopoiesis." *Nature Genetics* 44 (11): 1179–81.

Butler, P. G., A. D. Wanamaker Jr., J. D. Scourse, C. A. Richardson, and D. J. Reynolds. 2013. "Variability of Marine Climate on the North Icelandic Shelf in a 1357-Year Proxy Archive Based on Growth Increments in the Bivalve *Arctica islandica*." *Palaeogeography, Palaeoclimatology, Palaeoecology* 373: 141–51.

Cairns, J. 1975. "Mutation Selection and the Natural History of Cancer." *Nature* 255 (5505): 197–200.

Calle, E. E., and R. Kaaks. 2004. "Overweight, Obesity and Cancer: Epidemiological Evidence and Proposed Mechanisms." *Nature Reviews Cancer* 4 (8): 579–91.

Cancer Research UK. n.d. "Cancer Incidence by Age." Accessed August 6, 2017. http://www.cancerresearchuk.org/health-professional/cancer-stat istics/incidence/age#heading-Zero.

Carroll, S. B. 2006. *The Making of the Fittest: DNA and the Ultimate Forensic Record of Evolution.* New York: W. W. Norton.

Caspari, R., and S. H. Lee. 2004. "Older Age Becomes Common Late in Human Evolution." *Proceedings of the National Academy of Sciences of the United States of America* 101 (30): 10895–900.

Catlin, S. N., L. Busque, R. E. Gale, P. Guttorp, and J. L. Abkowitz. 2011. "The Replication Rate of Human Hematopoietic Stem Cells In Vivo." *Blood* 117 (17): 4460–66.

Caulin, A. F., T. A. Graham, L. Wang, and C. C. Maley. 2015. "Solutions to Peto's Paradox Revealed by Mathematical Modelling and Cross-Species Cancer Gene Analysis." *Philosophical Transactions of the Royal Society B: Biological Sciences* 370 (1673). Accessed September 3, 2017. http://rstb.royal societypublishing.org/content/royptb/370/1673/20140222.full.pdf.

Caulin, A. F., and C. C. Maley. 2011. "Peto's Paradox: Evolution's Prescription for Cancer Prevention." *Trends in Ecology and Evolution* 26 (4): 175–82.

Centers for Disease Control and Prevention. 2014. National Vital Statistics Reports 65, Number 4. https://www.cdc.gov/nchs/data/nvsr/nvsr65 /nvsr65_04.pdf.

Cha, R. S., W. G. Thilly, and H. Zarbl. 1994. "N-Nitroso-N-Methylurea-Induced Rat Mammary Tumors Arise from Cells with Preexisting Oncogenic HRAS1 Gene Mutations." *Proceedings of the National Academy of Sciences in the United States of America* 91 (9): 3749–53.

Chang, J., Y. Wang, L. Shao, R. Laberge, M. Demaria, J. Campisi, K. Janakiraman, et al. 2015. "Clearance of Senescent Cells by ABT263 Rejuvenates Aged Hematopoietic Stem Cells in Mice." *Nature Medicine* 22 (1): 78–83.

Chen, J., K. Sprouffske, Q. Huang, and C. C. Maley. 2011. "Solving the Puzzle of Metastasis: The Evolution of Cell Migration in Neoplasms." *PLoS One* 6 (4): e17933.

Claveria, C., G. Giovinazzo, S. Rocio, and M. Torres. 2013. "MYC-Driven Endogenous Cell Competition in the Early Mammalian Embryo." *Nature* 500 (7460): 39–44.

Cobrinik, D. 2013. "Learning about Retinoblastoma from Mouse Models That Missed." In *Animal Models of Brain Tumors*, edited by R. Martínez Murillo and A. Martínez, 141–52. Totowa, NJ: Humana Press.

Condeelis, J., and J. W. Pollard. 2006. "Macrophages: Obligate Partners for Tumor Cell Migration, Invasion, and Metastasis." *Cell* 124 (2): 263–66.

Coppé, J. P., P. Y. Desprez, A. Krtolica, and J. Campisi. 2010. "The Senescence-Associated Secretory Phenotype: The Dark Side of Tumor Suppression." *Annual Review of Pathology* 5: 99–118.

Corces-Zimmerman, M. R., and R. Majeti. 2014. "Pre-Leukemic Evolution of Hematopoietic Stem Cells—The Importance of Early Mutations in Leukemogenesis." *Leukemia* 28 (12): 2276–82.

Cornaro, A. 2014. *Writings on the Sober Life: The Art and Grace of Living Long.* Edited and translated by H. Fudemoto. Toronto: University of Toronto Press.

Cortopassi, G. A., and E. Wang. 1996. "There Is Substantial Agreement among Interspecies Estimates of DNA Repair Activity." *Mechanisms of Ageing and Development* 91 (3): 211–18.

Couraud, S., G. Zalcman, B. Milleron, F. Morin, and P. Souquet. 2012. "Lung Cancer in Never Smokers—A Review." *European Journal of Cancer* 48 (9): 1299–311.

Crawford, Y. G., M. L. Gauthier, A. Joubel, K. Mantei, K. Kozakiewicz, C. A. Afshari, and T. D. Tlsty. 2004. "Histologically Normal Human Mammary Epithelia with Silenced p16(INK4a) Overexpress COX-2, Promoting a Premalignant Program." *Cancer Cell* 5 (3): 263–73.

Crespi, B., and K. Summers. 2005. "Evolutionary Biology of Cancer." *Trends in Ecology and Evolution* 20 (10): 545–52.

Curtis, R. E., P. A. Rowlings, H. J. Deeg, D. A. Shriner, G. Socié, L. B. Travis, M. M. Horowitz, et al. 1997. "Solid Cancers after Bone Marrow Transplantation." *New England Journal of Medicine* 336 (13): 897–904.

D'Angelo, M. A., M. Raices, S. H. Panowski, and M. W. Hetzer. 2009. "Age-Dependent Deterioration of Nuclear Pore Complexes Causes a Loss of Nuclear Integrity in Post-Mitotic Cells." *Cell* 136 (2): 284–95.

Dantzer, B., and Q. E. Fletcher. 2015. "Telomeres Shorten More Slowly in Slow-Aging Wild Animals Than in Fast-Aging Ones." *Experimental Gerontology* 71: 38–47.

Daoust, S. P., L. Fahrig, A. E. Martin, and F. Thomas. 2013. "From Forest and Agro-Ecosystems to the Microecosystems of the Human Body: What Can Landscape Ecology Tell Us about Tumor Growth, Metastasis, and Treatment Options?" *Evolutionary Applications* 6 (1): 82–91.

Darwin, C. 1876. *The Origin of Species by Means of Natural Selection, Or The Preservation of Favoured Races in the Struggle for Life.* 6th ed. London: John Murray.

DeGregori, J. 2011. "Evolved Tumor Suppression: Why Are We So Good at Not Getting Cancer?" *Cancer Research* 71 (11): 3739–44.

——. 2012. "Challenging the Axiom: Does the Occurrence of Oncogenic Mutations Truly Limit Cancer Development with Age?" *Oncogene* 32 (15): 1869–75.

Deininger, M., E. Buchdunger, and B. J. Druker. 2005. "The Development of Imatinib as a Therapeutic Agent for Chronic Myeloid Leukemia." *Blood* 105 (7): 2640–53.

Demaria, M., N. Ohtani, S. A. Youssef, F. Rodier, W. Toussaint, J. R. Mitchell, R. Laberge, et al. 2014. "An Essential Role for Senescent Cells in Optimal Wound Healing through Secretion of PDGF-AA." *Developmental Cell* 31 (6): 722–33.

Demaria, M., M. N. O'Leary, J. Chang, L. Shao, S. Liu, F. Alimirah, K. Koenig, et al. 2016. "Cellular Senescence Promotes Adverse Effects of Chemotherapy and Cancer Relapse." *Cancer Discovery* 7 (2): 165–76.

Denic, A., R. J. Glassock, and A. D. Rule. 2016. "Structural and Functional Changes with the Aging Kidney." *Advances in Chronic Kidney Disease* 23 (1): 19–28.

De Pergola, G., and F. Silvestris. 2013. "Obesity as a Major Risk Factor for Cancer." *Journal of Obesity* 2013: 291546. Accessed September 3, 2017. https://www.hindawi.com/journals/jobe/2013/291546/.

DePinho, R. A. 2000. "The Age of Cancer." *Nature* 408 (6809): 248–54.

Dobzhansky, T. 1937. *Genetics and the Origin of Species.* New York: Columbia University Press.

——. 1973. "Nothing in Biology Makes Sense Except in the Light of Evolution." *American Biology Teacher* 35 (3): 125–29.

Domazet-Lošo, T. and D. Tautz. 2010. "Phylostratigraphic Tracking of Cancer Genes Suggests a Link to the Emergence of Multicellularity in Metazoa." *BMC Biology* 8: 66.

Dong, X., B. Milholland, and J. Vijg. 2016. "Evidence for a Limit to Human Lifespan." *Nature* 538 (7624): 257–59.

Druker, B. J. 2008. "Translation of the Philadelphia Chromosome into Therapy for CML." *Blood* 112 (13): 4808–17.

Du, Q., Y. Kawabe, C. Schilde, Z. Chen, and P. Schaap. 2015. "The Evolution of Aggregative Multicellularity and Cell-Cell Communication in the Dictyostelia." *Journal of Molecular Biology* 427 (23): 3722–33.

Eldredge, N. 1995. *Reinventing Darwin: The Great Debate at the High Table of Evolutionary Theory.* New York: Wiley.

——. 1999. *The Pattern of Evolution.* New York: W. H. Freeman and Co.

——, and S. J. Gould. 1972. "Punctuated Equilibria: An Alternative to Phyletic Gradualism." In *Models in Paleobiology,* edited by T. J. M. Schopf, 82–115. San Francisco: Freeman, Cooper, and Co.

Eller, E., J. Hawks, and J. H. Relethford. 2004. "Local Extinction and Recolonization, Species Effective Population Size, and Modern Human Origins." *Human Biology* 76 (5): 689–709.

Enciso-Mora, V., F. J. Hosking, E. Sheridan, S. E. Kinsey, T. Lightfoot, E. Roman, J. A. Irving, et al. 2012. "Common Genetic Variation Contributes Significantly to the Risk of Childhood B-Cell Precursor Acute Lymphoblastic Leukemia." *Leukemia* 26 (10): 2212–15.

Enriquez-Navas, P. M., Y. Kam, T. Das, S. Hassan, A. Silva, P. Foroutan, E. Ruiz, et al. 2016. "Exploiting Evolutionary Principles to Prolong Tumor Control in Preclinical Models of Breast Cancer." *Science Translational Medicine* 8 (327): 327ra24.

Ergen, A. V., and M. A. Goodell. 2010. "Mechanisms of Hematopoietic Stem Cell Aging." *Experimental Gerontology* 45 (4): 286–90.

Ewald, P. W., and H. A. Swain Ewald. 2013. "Toward a General Evolutionary Theory of Oncogenesis." *Evolutionary Applications* 6 (1): 70–81.

Fernandes, J. V., T. A. A. de Medeiros Fernandes, J. C. V. de Azevedo, R. N. O. Cobucci, M. G. F. de Carvalho, V. S. Andrade, and J. M. G. de Araújo. 2015. "Link between Chronic Inflammation and Human Papillomavirus-Induced Carcinogenesis (Review)." *Oncology Letters* 9 (3): 1015–26.

Fisher, R. A. 1930. *The Genetical Theory of Natural Selection*. Oxford: Clarendon Press of Oxford University Press.

Fleenor, C. J., K. Higa, M. M. Weil, and J. DeGregori. 2015. "Evolved Cellular Mechanisms to Respond to Genotoxic Insults: Implications for Radiation-Induced Hematologic Malignancies." *Radiation Research* 184 (4): 341–51.

Fleenor, C. J., A. I. Rozhok, V. Zaberezhnyy, D. Mathew, J. Kim, A. Tan, I. D. Bernstein, and J. DeGregori. 2015. "Contrasting Roles for C/EBPα and Notch in Irradiation-Induced Multipotent Hematopoietic Progenitor Cell Defects." *Stem Cells* 33 (4): 1345–58.

Flurkey, K., J. Papaconstantinou, and D. E. Harrison. 2002. "The Snell Dwarf Mutation Pit1(Dw) Can Increase Life Span in Mice." *Mechanisms of Ageing and Development* 123 (2–3): 121–30.

Folkman, J., and R. Kalluri. 2004. "Cancer without Disease." *Nature* 427 (6977): 787.

Fontana, L., L. Partridge, and V. D. Longo. 2010. "Extending Healthy Life Span—From Yeast to Humans." *Science* 328 (5976): 321–26.

Ford, A. M., C. Palmi, C. Bueno, D. Hong, P. Cardus, D. Knight, G. Cazzaniga, T. Enver, and M. Greaves. 2009. "The TEL-AML1 Leukemia Fusion Gene Dysregulates the TGF-beta Pathway in Early B Lineage Progenitor Cells." *Journal of Clinical Investigation* 119 (4): 826–36.

Fortunato, A., A. Boddy, D. Mallo, A. Aktipis, C. C. Maley, and J. W. Pepper. 2016. "Natural Selection in Cancer Biology: From Molecular Snowflakes to Trait Hallmarks." *Cold Spring Harbor Perspectives in Medicine* 7 (2): a029652.

Frank, S. A. 2010. "Somatic Evolutionary Genomics: Mutations during Development Cause Highly Variable Genetic Mosaicism with Risk of Cancer and Neurodegeneration." *Proceedings of the National Academy of Sciences of the United States of America* 107 (Supplement 1): 1725–30.

Galhardo, R. S., P. J. Hastings, and S. M. Rosenberg. 2007. "Mutation as a Stress Response and the Regulation of Evolvability." *Critical Reviews in Biochemistry and Molecular Biology* 42 (5): 399–435.

Garland, S. M., S. K. Kjaer, N. Muñoz, S. L. Block, D. R. Brown, M. J. DiNubile, B. R. Lindsay, et al. 2016. "Impact and Effectiveness of the Quadrivalent Human Papillomavirus Vaccine: A Systematic Review of 10 Years of Real-World Experience." *Clinical Infectious Diseases* 63 (4): 519–27.

Gatenby, R. A., and R. J. Gillies. 2008. "A Microenvironmental Model of Carcinogenesis." *Nature Reviews Cancer* 8 (1): 56–61.

Gatenby, R. A., A. S. Silva, R. J. Gillies, and B. R. Frieden. 2009. "Adaptive Therapy." *Cancer Research* 69 (11): 4894–903.

Genovese, G., A. K. Kähler, R. E. Handsaker, J. Lindberg, S. A. Rose, S. F. Bakhoum, K. Chambert, et al. 2014. "Clonal Hematopoiesis and Blood-Cancer Risk Inferred from Blood DNA Sequence." *New England Journal of Medicine* 371 (26): 2477–87.

Gerlinger, M., S. Horswell, J. Larkin, A. J. Rowan, M. P. Salm, I. Varela, R. Fisher, et al. 2014. "Genomic Architecture and Evolution of Clear Cell Renal Cell Carcinomas Defined by Multiregion Sequencing." *Nature Genetics* 46 (3): 225–33.

Godley, L. A., and R. A. Larson. 2008. "Therapy-Related Myeloid Leukemia." *Seminars in Oncology* 35 (4): 418–29.

Goldberg, A. D., C. D. Allis, and E. Bernstein. 2007. "Epigenetics: A Landscape Takes Shape." *Cell* 128 (4): 635–38.

Gomes, N. M. V., O. A. Ryder, M. L. Houck, S. J. Charter, W. Walker, N. R. Forsyth, S. N. Austad, et al. 2011. "Comparative Biology of Mammalian Telomeres: Hypotheses on Ancestral States and the Roles of Telomeres in Longevity Determination." *Aging Cell* 10 (5): 761–68.

Goodell, M. A., K. Brose, G. Paradis, A. S. Conner, and R. C. Mulligan. 1996. "Isolation and Functional Properties of Murine Hematopoietic Stem Cells That Are Replicating in Vivo." *Journal of Experimental Medicine* 183 (4): 1797–806.

Goodell, M. A., H. Nguyen, and N. Shroyer. 2015. "Somatic stem cell heterogeneity: diversity in the Blood, Skin and Intestinal Stem Cell Compartments." *Nature Reviews Molecular Cell Biology* 16 (5): 299–309.

Gorban, A. N., L. I. Pokidysheva, E. V. Smirnova, and T. A. Tyukina. 2011. "Law of the Minimum Paradoxes." *Bulletin of Mathematical Biology* 73 (9): 2013–44.

Gorre, M. E., and C. L. Sawyers. 2002. "Molecular Mechanisms of Resistance to STI571 in Chronic Myeloid Leukemia." *Current Opinion in Hematology* 9 (4): 303–7.

Goto, M. 2008. "Inflammaging (Inflammation + Aging): A Driving Force for Human Aging Based on an Evolutionarily Antagonistic Pleiotropy Theory?" *Bioscience Trends* 2 (6): 218–30.

Greaves, M. 2006. "Infection, Immune Responses and the Aetiology of Childhood Leukaemia." *Nature Reviews Cancer* 6 (3): 193–203.

———. 2014. "Was Skin Cancer a Selective Force for Black Pigmentation in Early Hominin Evolution?" *Proceedings of the Royal Society B: Biological Sciences* 281 (1781): 20132955.

Green, D. R., and B. Levine. 2014. "To Be or Not to Be? How Selective Autophagy and Cell Death Govern Cell Fate." *Cell* 157 (1): 65–75.

Grywalska, E., and J. Rolinski. 2015. "Epstein-Barr Virus–Associated Lymphomas." *Seminars in Oncology* 42 (2): 291–303.

Gunes, C., and K. L. Rudolph. 2013. "The Role of Telomeres in Stem Cells and Cancer." *Cell* 152 (3): 390–3.

Gurven, M., and H. Kaplan. 2007. "Longevity among Hunter-Gatherers: A Cross-Cultural Examination." *Population and Development Review* 33 (2): 321–65.

Haddow, A. 1938. "Cellular Inhibition and the Origin of Cancer." *Acta Unio Internationalis Contra Cancrum* 3: 342–52.

Hainaut, P., and G. P. Pfeifer. 2001. "Patterns of p53 G → T Transversions in Lung Cancers Reflect the Primary Mutagenic Signature of DNA-Damage by Tobacco Smoke." *Carcinogenesis* 22 (3): 367–74.

Haldane, J. B. S. 1932. *The Causes of Evolution*. London: Longmans, Green and Co.

Hamilton, W. D. 1966. "The Moulding of Senescence by Natural Selection." *Journal of Theoretical Biology* 12 (1): 12–45.

Hammond, E. M., and A. J. Giaccia. 2005. "The Role of p53 in Hypoxia-Induced Apoptosis." *Biochemical and Biophysical Research Communications* 331 (3): 718–25.

Hanahan, D., and R. A. Weinberg. 2000. "The Hallmarks of Cancer." *Cell* 100 (1): 57–70.

———. 2011. "Hallmarks of Cancer: The Next Generation." *Cell* 144 (5): 646–74.

Harris, S. S. 2006. "Vitamin D and African Americans." *Journal of Nutrition* 136 (4): 1126–29.

Haussmann, M. F., D. W. Winkler, K. M. O'Reilly, C. E. Huntington, I. C. T. Nisbet, and C. M. Vleck. 2003. "Telomeres Shorten More Slowly in Long-Lived Birds and Mammals Than in Short-Lived Ones." *Proceedings of the Royal Society of London. Series B: Biological Sciences* 270 (1522): 1387–92.

Hecht, S. S. 1999. "Tobacco Smoke Carcinogens and Lung Cancer." *Journal of the National Cancer Institute* 91 (14): 1194–210.

Hemminki, K., and Y. Jiang. 2002. "Risks among Siblings and Twins for Childhood Acute Lymphoid Leukaemia: Results from the Swedish Family-Cancer Database." *Leukemia* 16 (2): 297–98.

Henry, C. J., M. Casás-Selves, J. Kim, V. Zaberezhnyy, L. Aghili, A. E. Daniel, L. Jimenez, et al. 2015. "Aging-Associated Inflammation Promotes Selection for Adaptive Oncogenic Events in B Cell Progenitors." *Journal of Clinical Investigation* 125 (12): 4666–80.

Henry, C. J., A. Marusyk, and J. DeGregori. 2011. "Aging-Associated Changes in Hematopoiesis and Leukemogenesis: What's the Connection?" *Aging* 3 (6): 643–56.

Henry, C. J., A. Marusyk, V. Zaberezhnyy, B. Adane, and J. DeGregori. 2010. "Declining Lymphoid Progenitor Fitness Promotes Aging-Associated Leukemogenesis." *Proceedings of the National Academy of Sciences of the United States of America* 107 (50): 21713–18.

Herbst, R. S., and P. A. Bunn Jr. 2003. "Targeting the Epidermal Growth Factor Receptor in Non-Small Cell Lung Cancer." *Clinical Cancer Research* 9 (16 Part 1): 5813–24.

Hietpas, R. T., C. Bank, J. D. Jensen, and D. N. A. Bolon. 2013. "Shifting Fitness Landscapes in Response to Altered Environments." *Evolution* 67 (12): 3512–22.

Hirsch, F. R., K. Suda, J. Wiens, and P. A. Bunn Jr. 2016. "New and Emerging Targeted Treatments in Advanced Non-Small-Cell Lung Cancer." *Lancet* 388 (10048): 1012–24.

Hochberg, M. E., and R. J. Noble. 2017. "A Framework for How Environment Contributes to Cancer Risk." *Ecology Letters* 20 (2): 117–34.

Hoeijmakers, J. H. 2001. "Genome Maintenance Mechanisms for Preventing Cancer." *Nature* 411 (6835): 366–74.

———. 2009. "DNA Damage, Aging, and Cancer." *New England Journal of Medicine* 361 (15): 1475–85.

Hoelzl, F., S. Smith, J. S. Cornils, D. Aydinonat, C. Bieber, and T. Ruf. 2016. "Telomeres Are Elongated in Older Individuals in a Hibernating Rodent, the Edible Dormouse (*Glis glis*)." *Scientific Reports* 6: 36856.

Horvath, S. 2013. "DNA Methylation Age of Human Tissues and Cell Types." *Genome Biology* 14 (10): R115.

Hunger, S. P., and C. G. Mullighan. 2015. "Acute Lymphoblastic Leukemia in Children." *New England Journal of Medicine* 373 (16): 1541–52

Hursting, S. D., S. M. Dunlap, N. A. Ford, M. J. Hursting, and L. M. Lashinger. 2013. "Calorie Restriction and Cancer Prevention: A Mechanistic Perspective." *Cancer and Metabolism* 1 (1): 10.

Ibrahim-Hashim, A., M. Robertson-Tessi, P. M. Enriquez-Navas, M. Damaghi, Y. Balagurunathan, J. W. Wojtkowiak, S. Russell, et al. 2017. "Defining Cancer Subpopulations by Adaptive Strategies Rather Than Molecular Properties Provides Novel Insights into Intratumoral Evolution." *Cancer Research* 77 (9): 2242–54.

Inoue, S., W. Y. Li, A. Tseng, I. Beerman, A. J. Elia, S. C. Bendall, F. Lemonnier, et al. 2016. "Mutant IDH1 Downregulates ATM and Alters DNA Repair and Sensitivity to DNA Damage Independent of TET2." *Cancer Cell* 30 (2): 337–48.

Irfan, U. 2013. "Climate Change May Have Spurred Human Evolution." *Scientific American*, January 2. Accessed August 6, 2017. https://www.scientificamerican.com/article/climate-change-may-have-spurred-human-evolution/.

Islami, F., P. Boffetta, J. Ren, L. Pedoeim, D. Khatib, and F. Kamangar. 2009. "High-Temperature Beverages and Foods and Esophageal Cancer Risk—A Systematic Review." *International Journal of Cancer* 125 (3): 491–524.

Ivory, S. J., M. W. Blome, J. W. King, M. M. McGlue, J. E. Cole, and A. S. Cohen. 2016. "Environmental Change Explains Cichlid Adaptive Radiation at Lake Malawi over the Past 1.2 Million Years." *Proceedings of the National Academy of Sciences of the United States of America* 113 (42): 11895–900.

Iwasaki, A., and R. Medzhitov. 2015. "Control of Adaptive Immunity by the Innate Immune System." *Nature Immunology* 16 (4): 343–53.

Jablonski, N. G., and G. Chaplin. 2010. "Human Skin Pigmentation as an Adaptation to UV Radiation." *Proceedings of the National Academy of Sciences of the United States of America* 107 (Supplement 2): 8962–68.

Jackson, J. A., I. M. Friberg, S. Little, and J. E. Bradley. 2009. "Review Series on Helminths, Immune Modulation and the Hygiene Hypothesis: Immunity against Helminths and Immunological Phenomena in Modern Human Populations: Coevolutionary Legacies?" *Immunology* 126 (1): 18–27.

Jaiswal, S., P. Fontanillas, J. Flannick, A. Manning, P. V. Grauman, B. G. Mar, R. C. Lindsley, et al. 2014. "Age-Related Clonal Hematopoiesis Associated with Adverse Outcomes." *New England Journal of Medicine* 371 (26): 2488–98.

Janeway, C. A., Jr., and R. Medzhitov. 2002. "Innate Immune Recognition." *Annual Review of Immunology* 20: 197–216.

Jonason, A. S., S. Kunala, G. J. Price, R. J. Restifo, H. M. Spinelli, J. A. Persing, D. J. Leffell, R. E. Tarone, and D. E. Brash. 1996. "Frequent Clones of p53-Mutated Keratinocytes in Normal Human Skin." *Proceedings of the National Academy of Sciences of the United States of America* 93 (24): 14025–29.

Kajita, M., and Y. Fujita. 2015. "EDAC: Epithelial Defence against Cancer-Cell Competition between Normal and Transformed Epithelial Cells in Mammals." *Journal of Biochemistry* 158 (1): 15–23.

Kaletta, T., and M. O. Hengartner. 2006. "Finding Function in Novel Targets: C. Elegans as a Model Organism." *Nature Reviews Drug Discovery* 5 (5): 387–99.

Kam, Y., T. Das, H. Tian, P. Foroutan, E. Ruiz, G. Martinez, S. Minton, R. J. Gillies, and R. A. Gatenby. 2015. "Sweat but No Gain: Inhibiting Proliferation of Multidrug Resistant Cancer Cells with 'Ersatzdroges.'" *International Journal of Cancer* 136 (4): E188–96.

Kandoth, C., M. D. McLellan, F. Vandin, K. Ye, B. Niu, C. Lu, M. Xie, et al. 2013. "Mutational Landscape and Significance across 12 Major Cancer Types." *Nature* 502 (7471): 333–39.

Kang, T. W., T. Yevsa, N. Woller, L. Hoenicke, T. Wuestefeld, D. Dauch, A. Hohmeyer, et al. 2011. "Senescence Surveillance of Pre-Malignant Hepatocytes Limits Liver Cancer Development." *Nature* 479 (7374): 547–51.

Kaplan, H., K. Hill, J. Lancaster, and A. M. Hurtado. 2000. "A Theory of Human Life History Evolution: Diet, Intelligence, and Longevity." *Evolutionary Anthropology* 9 (4): 156–85.

Kaplan, H. S., and M. B. Brown. 1951. "Further Observations on Inhibition of Lymphoid Tumor Development by Shielding and Partial-Body Irradiation of Mice." *Cancer Research* 12 (2): 427–36.

Karachaliou, N., S. Pilotto, C. Lazzari, E. Bria, F. de Marinis, and R. Rosell. 2016. "Cellular and Molecular Biology of Small Cell Lung Cancer: An Overview." *Translational Lung Cancer Research* 5 (1): 2–15.

Kenyon, C. J. 2010. "The Genetics of Ageing." *Nature* 464 (7288): 504–12.

Khalil, D. N., E. L. Smith, R. J. Brentjens, and J. D. Wolchok. 2016. "The Future of Cancer Treatment: Immunomodulation, CARS and Combination Immunotherapy." *Nature Reviews Clinical Oncology* 13 (5): 273–90.

Kim, C. F. B., E. L. Jackson, D. G. Kirsch, J. Grimm, A. T. Shaw, K. Lane, J. Kissil, et al. 2005. "Mouse Models of Human Non-Small-Cell Lung Cancer: Raising the Bar." *Cold Spring Harbor Symposia on Quantitative Biology* 70: 241–50.

Kinlen, L. 1988. Evidence for an Infective Cause of Childhood Leukaemia: Comparison of a Scottish New Town with Nuclear Reprocessing Sites in Britain. *Lancet* 2 (8624): 1323–27.

Kirkwood, T. B. 2005. "Understanding the Odd Science of Aging." *Cell* 120 (4): 437–47.

Kominami, R., and O. Niwa. 2006. "Radiation Carcinogenesis in Mouse Thymic Lymphomas." *Cancer Science* 97 (7): 575–81.

Kostadinov, R. L., M. K. Kuhner, X. Li, C. A. Sanchez, P. C. Galipeau, T. G. Paulson, C. L. Sather, et al. 2013. "NSAIDs Modulate Clonal Evolution in Barrett's Esophagus." *PLoS Genetics* 9 (6): e1003553.

Kotas, M. E,. and R. Medzhitov. 2015. "Homeostasis, Inflammation, and Disease Susceptibility." *Cell* 160 (5): 816–27.

Kowald, A., and T. B. L. Kirkwood. 2015. "Evolutionary Significance of Ageing in the Wild." *Experimental Gerontology* 71: 89–94.

Kruger, P., M. Saffarzadeh, A. N. R. Weber, N. Rieber, M. Radsak, H. von Bernuth, C. Benarafa, D. Roos, J. Skokowa, and D. Hartl. 2015. "Neutrophils: Between Host Defence, Immune Modulation, and Tissue Injury." *PLoS Pathogens* 11 (3): e1004651.

Kuhn, T. S. 1962. *The Structure of Scientific Revolutions*. Chicago: University of Chicago Press.

Kvell, K., J. Pongrácz, M. Székely, M. Balaskó, E. Pétervári, and G. Bakó. 2011. "Gerontology." In *Molecular and Clinical Basics of Gerontology*, edited by J. Pongrácz, M. Székely, M. Balaskó, G. Bakó, and R. Bognár. Pécs, Hungary: University of Pécs. Accessed September 3, 2017. http://www.tankonyvtar.hu /hu/tartalom/tamop425/0011_1A_Gerontologia_en_book/index.html.

Kwiatkowski, F., M. Arbre, Y. Bidet, C. Laquet, N. Uhrhammer, and Y. J. Bignon. 2015. "BRCA Mutations Increase Fertility in Families at Hereditary Breast / Ovarian Cancer Risk." *PLoS One* 10(6): e0127363.

Laayouni, H., M. Oosting, P. Luisi, M. Ioana, S. Alonso, I. Ricaño-Ponce, G. Trynka, et al. 2014. "Convergent Evolution in European and Rroma Populations Reveals Pressure Exerted by Plague on Toll-Like Receptors." *Proceedings of the National Academy of Sciences of the United States of America* 111 (7): 2668–73.

Lamming, D. W., L. Ye, P. Katajisto, M. D. Goncalves, M. Saitoh, D. M. Stevens, J. G. Davis, et al. 2012. "Rapamycin-Induced Insulin Resistance Is Mediated by mTORC2 Loss and Uncoupled from Longevity." *Science* 335 (6076): 1638–43.

Land, H., L. F. Parada, and R. A. Weinberg. 1983. "Tumorigenic Conversion of Primary Embryo Fibroblasts Requires at Least Two Cooperating Oncogenes." *Nature* 304 (5927): 596–602.

Layde, P. M., L. A. Webster, A. L. Baughman, P. A. Wingo, G. L. Rubin, and H. W. Ory. 1989. "The Independent Associations of Parity, Age at First Full Term Pregnancy, and Duration of Breastfeeding with the Risk of Breast Cancer." *Journal of Clinical Epidemiology* 42 (10): 963–73.

Le Couteur, D. G., S. Solon-Biet, V. C. Cogger, S. J. Mitchell, A. Senior, R. de Cabo, D. Raubenheimer, and S. J. Simpson. 2016. "The Impact of Low-Protein High-Carbohydrate Diets on Aging and Lifespan." *Cellular and Molecular Life Sciences* 73 (6): 1237–52.

Lensch, M. W., R. K. Rathbun, S. B. Olson, G. R. Jones, and G. C. Bagby Jr. 1999. "Selective Pressure as an Essential Force in Molecular Evolution of Myeloid Leukemic Clones: A View from the Window of Fanconi Anemia." *Leukemia* 13 (11): 1784–89.

Leroi, A. M., V. Koufopanou, and A. Burt. 2003. "Cancer Selection." *Nature Reviews Cancer* 3 (3): 226–31.

Lescale, C., S. Dias, J. Maës, A. Cumano, P. Szabo, D. Charron, M. E. Weksler, C. Dosquet, P. Vieira, and M. Goodhardt. 2010. "Reduced EBF Expression Underlies Loss of B Cell Potential of Hematopoietic Progenitors with Age." *Aging Cell* 9 (3): 410–19.

Le Tourneau, C., J. J. Lee, and L. L. Siu. 2009. "Dose Escalation Methods in Phase I Cancer Clinical Trials." *Journal of the National Cancer Institute* 101(10): 708–720.

Levine, A. J. 2009. "The Common Mechanisms of Transformation by the Small DNA Tumor Viruses: The Inactivation of Tumor Suppressor Gene Products: p53." *Virology* 384 (2): 285–93.

Levrero, M. 2006. "Viral Hepatitis and Liver Cancer: The Case of Hepatitis C." *Oncogene* 25 (27): 3834–47.

Li, J., D. P. Sejas, X. Zhang, Y. Qiu, K. J. Nattamai, R. Rani, K. R. Rathbun, et al. 2007. "TNF-Alpha Induces Leukemic Clonal Evolution Ex Vivo in Fanconi Anemia Group C Murine Stem Cells." *Journal of Clinical Investigation* 117 (11): 3283–95.

Lin, M. V., L. Y. King, and R. T. Chung. 2015. "Hepatitis C Virus-Associated Cancer." *Annual Review of Pathology* 10: 345–70.

Lindblad-Toh, K., M. Garber, O. Zuk, M. F. Lin, B. J. Parker, S. Washietl, P. Kheradpour, et al. 2011. "A High-Resolution Map of Human Evolutionary Constraint Using 29 Mammals." *Nature* 478 (7370): 476–82.

Lipinski, K. A., L. J. Barber, M. N. Davies, M. Ashenden, A. Sottoriva, and M. Gerlinger. 2016. "Cancer Evolution and the Limits of Predictability in Precision Cancer Medicine." *Trends in Cancer* 2 (1): 49–63.

Lisanti, M. P., U. E. Martinez-Outschoorn, S. Pavlides, D. Whitaker-Menezes, R. G. Pestell, A. Howell, and F. Sotgia. 2011. "Accelerated Aging in the

Tumor Microenvironment: Connecting Aging, Inflammation and Cancer Metabolism with Personalized Medicine." *Cell Cycle* 10 (13): 2059–63.

Little, J. B. 2000. "Radiation Carcinogenesis." *Carcinogenesis* 21 (3): 397–404.

Lombard, D. B., K. F. Chua, R. Mostoslavsky, S. Franco, M. Gostissa, and F. W. Alt. 2005. "DNA Repair, Genome Stability, and Aging." *Cell* 120 (4): 497–512.

Lopez-Garcia, C., A. M. Klein, B. D. Simons, and D. J. Winton. 2010. "Intestinal Stem Cell Replacement Follows a Pattern of Neutral Drift." *Science* 330 (6005): 822–25.

Lowe, S. W., E. Cepero, and G. Evan. 2004. "Intrinsic Tumour Suppression." *Nature* 432 (7015): 307–15.

Luebeck, E. G., and S. H. Moolgavkar. 2002. "Multistage Carcinogenesis and the Incidence of Colorectal Cancer." *Proceedings of the National Academy of Sciences of the United States of America* 99 (23): 15095–100.

Lynch, M. 2010. "Evolution of the Mutation Rate." *Trends in Genetics* 26 (8): 345–52.

Madeo, F., A. Zimmermann, M. C. Maiuri, and G. Kroemer. 2015. "Essential Role for Autophagy in Life Span Extension." *Journal of Clinical Investigation* 125 (1): 85–93.

Madsen, T., A. Arnal, M. Vittecoq, F. Bernex, J. Abadie, S. Labrut, D. Garcia, et al. 2017. "Cancer Prevalence and Etiology in Wild and Captive Animals." In *Ecology and Evolution of Cancer*, edited by B. Ujvari, B. Roche and F. Thomas, 11–46. London: Academic Press.

Maffini, M. V., A. M. Soto, J. M. Calabro, A. A. Ucci, and C. Sommenschein. 2004. "The Stroma as a Crucial Target in Rat Mammary Gland Carcinogenesis." *Journal of Cell Science* 117 (8): 1495–502.

Maley, C. C., B. J. Reid, and S. Forrest. 2004. "Cancer Prevention Strategies That Address the Evolutionary Dynamics of Neoplastic Cells: Simulating Benign Cell Boosters and Selection for Chemosensitivity." *Cancer Epidemiology, Biomarkers and Prevention* 13 (8): 1375–84.

Malinsky, M., and W. Salzburger. 2016. "Environmental Context for Understanding the Iconic Adaptive Radiation of Cichlid Fishes in Lake Malawi." *Proceedings of the National Academy of Sciences of the United States of America* 113 (42): 11654–56.

Malthus, T. R. 1798. *An Essay on the Principle of Population.* London: J. Johnson.

Martincorena, I., A. Roshan, M. Gerstung, P. Ellis, P. Van Loo, S. McLaren, D. C. Wedge, et al. 2015. "High Burden and Pervasive Positive Selection of Somatic Mutations in Normal Human Skin." *Science* 348 (6237): 880–86.

Marusyk, A., V. Almendro, and K. Polyak. 2012. "Intra-Tumour Heterogeneity: A Looking Glass for Cancer?" *Nature Reviews Cancer* 12 (5): 323–34.

Marusyk, A., M. Casás-Selves, C. J. Henry, V. Zaberezhnyy, J. Klawitter, U. Christians, and J. DeGregori. 2009. "Irradiation Alters Selection for Oncogenic Mutations in Hematopoietic Progenitors." *Cancer Research* 69 (18): 7262–69.

Marusyk, A., and J. DeGregori. 2008. "Declining Cellular Fitness with Age Promotes Cancer Initiation by Selecting for Adaptive Oncogenic Mutations." *Biochimica et Biophysica Acta* 1785 (1): 1–11.

Marusyk, A., C. C. Porter, V. Zaberezhnyy, and J. DeGregori. 2010. "Irradiation Selects for p53-Deficient Hematopoietic Progenitors." *PLoS Biology* 8 (3): e1000324.

Matioli, G. T. 2002. "BCR-ABL Insufficiency for the Transformation of Human Stem Cells into CML." *Medical Hypotheses* 59 (5): 588–89.

Mayr, E. 1982. *The Growth of Biological Thought: Diversity, Evolution, and Inheritance*. Cambridge, MA: Belknap Press of Harvard University Press.

Mazzone, M., D. Dettori, R. L. de Oliveira, S. Loges, T. Schmidt, B. Jonckx, Y. Tian, et al. 2009. "Heterozygous Deficiency of *PHD2* Restores Tumor Oxygenation and Inhibits Metastasis via Endothelial Normalization." *Cell* 136 (5): 839–51.

McCay, C. M., M. F. Crowell, L. A. Maynard. 1935. The Effect of Retarded Growth upon the Length of Life Span and upon the Ultimate Body Size. *The Journal of Nutrition* 10: 63–79.

McKerrell, T., N. Park, T. Moreno, C. S. Grove, H. Ponstingl, J. Stephens, Understanding Society Scientific Group, et al. 2015. "Leukemia-Associated Somatic Mutations Drive Distinct Patterns of Age-Related Clonal Hemopoiesis." *Cell Reports* 10 (8): 1239–45.

Medawar, P. 1952. *An Unsolved Problem of Biology*. London: H. K. Lewis and Co.

Melnyk, A. H., A. Wong, and R. Kassen. 2015. "The Fitness Costs of Antibiotic Resistance Mutations." *Evolutionary Applications* 8 (3): 273–83.

Menéndez, J., A. Pérez-Garijo, M. Calleja, and G. Morata. 2010. "A Tumor-Suppressing Mechanism in Drosophila Involving Cell Competition and the Hippo Pathway." *Proceedings of the National Academy of Sciences of the United States of America* 107 (33): 14651–56.

Mercken, E. M., B. A. Carboneau, S. M. Krzysik-Walker, and R. de Cabo. 2012. "Of Mice and Men: The Benefits of Caloric Restriction, Exercise, and Mimetics." *Ageing Research Reviews* 11 (3): 390–98.

Mintz, B., and K. Illmensee. 1975. "Normal Genetically Mosaic Mice Produced from Malignant Teratocarcinoma Cells." *Proceedings of the National Academy of Sciences of the United States of America* 72 (9): 3585–89.

Mirzaei, H., R. Raynes, and V. D. Longo. 2016. "The Conserved Role of Protein Restriction in Aging and Disease." *Current Opinion in Clinical Nutrition and Metabolic Care* 19 (1): 74–79.

Mittal, D., M. M. Gubin, R. D. Schreiber, and M. J. Smyth. 2014. "New Insights into Cancer Immunoediting and Its Three Component Phases—Elimination, Equilibrium and Escape." *Current Opinion in Immunology* 27: 16–25.

Miyajima, A., M. Tanaka, and T. Itoh. 2014. "Stem / Progenitor Cells in Liver Development, Homeostasis, Regeneration, and Reprogramming." *Cell Stem Cell* 14 (5): 561–74.

Mody, R., S. Li, D. C. Dover, S. Sallan, W. Leisenring, K. C. Oeffinger, Y. Yasui, L. L. Robison, and J. P. Neglia. 2008. "Twenty-Five-Year Follow-Up among Survivors of Childhood Acute Lymphoblastic Leukemia: A Report from the Childhood Cancer Survivor Study." *Blood* 111 (12): 5515–23.

Monod, J. 1972. *Chance and Necessity: An Essay on the Natural Philosophy of Modern Biology*. New York: Vintage Books.

Morata, G., and P. Ripoll. 1975. "Minutes: Mutants of Drosophila Autonomously Affecting Cell Division Rate." *Developmental Biology* 42 (2): 211–21.

Morrison, S. J., and D. T. Scadden. 2014. "The Bone Marrow Niche for Haematopoietic Stem Cells." *Nature* 505 (7483): 327–34.

Mortaz, E., P. Tabarsi, D. Mansouri, A. Khosravi, J. Garssen, A. Velayati, and I. M. Adcock. 2016. "Cancers Related to Immunodeficiencies: Update and Perspectives." *Frontiers in Immunology* 7: 365.

Morton, L. M., G. M. Dores, M. A. Tucker, C. J. Kim, K. Onel, E. S. Gilbert, J. F. Fraumeni Jr., and R. E. Curtis. 2013. "Evolving Risk of Therapy-Related Acute Myeloid Leukemia Following Cancer Chemotherapy among Adults in the United States, 1975–2008." *Blood* 121 (15): 2996–3004.

Mueller, S., A. Wahlander, N. Selevsek, C. Otto, E. M. Ngwa, K. Poljak, A. D. Frey, M. Aebi, and R. Gauss. 2015. "Protein Degradation Corrects for Imbalanced Subunit Stoichiometry in OST Complex Assembly." *Molecular Biology of the Cell* 26 (14): 2596–608.

Muezzinler, A., A. K. Zaineddin, and H. Brenner. 2013. "A systematic Review of Leukocyte Telomere Length and Age in Adults." *Ageing Research Reviews* 12 (2): 509–19.

Müller, A., and R. Fishel. 2002. "Mismatch Repair and the Hereditary Non-Polyposis Colorectal Cancer Syndrome (HNPCC)." *Cancer Investigation* 20 (1): 102–9.

Muller, H. J. 1948. "Evidence of the Precision of Genetic Adaptation." *Harvey Lecture Series* 43: 165–229.

Mullighan, C. G., C. B. Miller, I. Radtke, L. A. Phillips, J. Dalton, J. Ma, D. White, et al. 2008. "BCR-ABL1 Lymphoblastic Leukaemia Is Characterized by the Deletion of Ikaros." *Nature* 453 (7191): 110-14.

Mutter, G. L., T. A. Ince, J. P. A. Baak, G. A. Kust, X. Zhou, and C. Eng. 2001. "Molecular Identification of Latent Precancers in Histologically Normal Endometrium." *Cancer Research* 61 (11): 4311-14.

National Cancer Institute. 2014. "SEER*Stat Databases: November 2014 Submission." Accessed August 9, 2017. https://seer.cancer.gov/data /seerstat/nov2014/.

National Cancer Institute. 2016. "NCI Budget Fact Book Archive." Accessed September 2, 2017. https://www.cancer.gov/about-nci/budget/fact-book /archive.

Nesse, R. M., and G. C. Williams. 1996. *Evolution and Healing: The New Science of Darwinian Medicine*. London: Phoenix.

Nguyen, D. X., P. D. Bos, and J. Massague. 2009. "Metastasis: From Dissemination to Organ-Specific Colonization." *Nature Reviews Cancer* 9(4): 274-284.

Nielsen, J., R. B. Hedeholm, J. Heinemeier, P. G. Bushnell, J. S. Christiansen, J. Olsen, C. B. Ramsey, et al. 2016. "Eye Lens Radiocarbon Reveals Centuries of Longevity in the Greenland Shark (*Somniosus microcephalus*)." *Science* 353 (6300): 702-4.

Nilsson, J. A., and J. L. Cleveland. 2003. "MYC Pathways Provoking Cell Suicide and Cancer." *Oncogene* 22 (56): 9007-9021.

Nordling, C. O. 1953. "A New Theory on the Cancer-Inducing Mechanism." *British Journal of Cancer* 7 (1): 68-72.

Nowell, P., and D. Hungerford. 1960. "A Minute Chromosome in Human Chronic Granulocytic Leukemia." *Science* 132: 1497.

Nowell, P. C. 1976. "The Clonal Evolution of Tumor Cell Populations." *Science* 194 (4260): 23-28.

O'Callaghan, D. S., D. O'Donnell, F. O'Connell, and K. J. O'Byrne. 2010. "The Role of Inflammation in the Pathogenesis of Non-Small Cell Lung Cancer." *Journal of Thoracic Oncology* 5 (12): 2024-36.

Orr, H. A. 1998. "The Population Genetics of Adaptation: The Distribution of Factors Fixed during Adaptive Evolution." *Evolution* 52 (4): 935-49.

——. 2005. "The Genetic Theory of Adaptation: A Brief History." *Nature Reviews Genetics* 6 (2): 119-27.

Ou, J., F. Carbonero, E. G. Zoetendal, J. P. DeLany, M. Wang, K. Newton, H. R. Gaskins, and S. J. D. O'Keefe. 2013. "Diet, Microbiota, and Microbial Metabolites in Colon Cancer Risk in Rural Africans and African Americans." *American Journal of Clinical Nutrition* 98 (1): 111-20.

Paterson, C., M. A. Nowak, and B. Waclaw. 2016. "An Exactly Solvable, Spatial Model of Mutation Accumulation in Cancer." *Scientific Reports* 6: 39511.

Pelengaris, S., M. Khan, and G. Evan. 2002. "c-MYC: More Than Just a Matter of Life and Death." *Nature Reviews Cancer* 2 (10): 764–76.

Peto, R., F. J. Roe, P. N. Lee, L. Levy, and J. Clack. 1975. "Cancer and Ageing in Mice and Men." *British Journal of Cancer* 32 (4): 411–26.

Pfeifer, G. P., M. F. Denissenko, M. Olivier, N. Tretyakova, S. S. Hecht, and P. Hainaut. 2002. "Tobacco Smoke Carcinogens, DNA Damage and p53 Mutations in Smoking-Associated Cancers." *Oncogene* 21 (48): 7435–51.

Pickup, M. W., J. K. Mouw, and V. M. Weaver. 2014. "The Extracellular Matrix Modulates the Hallmarks of Cancer." *EMBO Reports* 15 (12): 1243–53.

Proctor, R. N. "The History of the Discovery of the Cigarette–Lung Cancer Link: Evidentiary Traditions, Corporate Denial, Global Toll." *Tobacco Control* 21 (2): 87–91.

Promislow, D. E. L. 1994. "DNA Repair and the Evolution of Longevity: A Critical Analysis." *Journal of Theoretical Biology* 170 (3): 291–300.

Pyo, J. O., S. M. Yoo, H. H. Ahn, J. Nah, S. H. Hong, T. I. Kam, S. Jung, and Y. K. Jung. 2013. "Overexpression of Atg5 in Mice Activates Autophagy and Extends Lifespan." *Nature Communications* 4: 2300.

Rangarajan, A., S. J. Hong, A. Gifford, and R. A. Weinberg. 2004. "Species- and Cell Type–Specific Requirements for Cellular Transformation." *Cancer Cell* 6 (2): 171–83.

Reya, T., S. J. Morrison, M. F. Clarke, and I. L. Weissman. 2001. "Stem Cells, Cancer, and Cancer Stem Cells." *Nature* 414 (6859): 105–11.

Ridker, P. M., J. G. MacFadyen, T. Thuren, B. M. Everett, P. Libby, R. J. Glynn, P. Ridker, et al. (2017). "Effect of interleukin-1β; inhibition with canakinumab on incident lung cancer in patients with atherosclerosis: exploratory results from a randomised, double-blind, placebo-controlled trial." *Lancet.* http://dx.doi.org/10.1016/S0140-6736(17)32247-X.

Roche, B., A. P. Møller, J. DeGregori, and F. Thomas. 2017. "Cancer in Animals: Reciprocal Feedbacks between Evolution of Cancer Resistance and Ecosystem Functioning." In *Ecology and Evolution of Cancer*, edited by B. Ujvari, B. Roche and F. Thomas, 181–91. London: Academic Press.

Rook, G. A. W., and A. Dalgleish. 2011. "Infection, Immunoregulation, and Cancer." *Immunological Reviews* 240 (1): 141–59.

Rooks, M. G., and W. S. Garrett. 2016. "Gut Microbiota, Metabolites and Host Immunity." *Nature Reviews Immunology* 16 (6): 341–52.

Rostom, A., C. Dube, and G. Lewin. 2007. "Use of Aspirin and NSAIDs to Prevent Colorectal Cancer." Rockville, MD: Agency for Healthcare Research and Quality.

Rothwell, P. M., F. G. Fowkes, J. F. Belch, H. Ogawa, C. P. Warlow, and T. W. Meade. 2011. "Effect of Daily Aspirin on Long-Term Risk of Death Due to Cancer: Analysis of Individual Patient Data from Randomised Trials." *Lancet* 377 (9759): 31–41.

Rowley, J. D. 1973. Chromosomal patterns in myelocytic leukemia. *The New England journal of medicine* 289 (4): 220–21.

Rozhok, A. I., and J. DeGregori. 2015. "Toward an Evolutionary Model of Cancer: Considering the Mechanisms That Govern the Fate of Somatic Mutations." *Proceedings of the National Academy of Sciences of the United States of America* 112 (29): 8914–21.

———. 2016. "The Evolution of Lifespan and Age-Dependent Cancer Risk." *Trends in Cancer* 2 (10): 552–60.

Rozhok, A. I., J. L. Salstrom, and J. DeGregori. 2014. "Stochastic Modeling Indicates That Aging and Somatic Evolution in the Hematopoietic System Are Driven by Non-Cell-Autonomous Processes." *Aging* 6 (12): 1033–48.

———. 2016. "Stochastic Modeling Reveals an Evolutionary Mechanism Underlying Elevated Rates of Childhood Leukemia." *Proceedings of the National Academy of Sciences of the United States of America* 113 (4): 1050–55.

Rubin, H. 2002. "Selective Clonal Expansion and Microenvironmental Permissiveness in Tobacco Carcinogenesis." *Oncogene* 21 (48): 7392–411.

———. 2007. "Ordered Heterogeneity and Its Decline in Cancer and Aging." *Advances in Cancer Research* 98: 117–47.

Rudolph, K. L., S. Chang, H. W. Lee, M. Blasco, G. J. Gottlieb, C. Greider, and R. A. DePinho. 1999. "Longevity, Stress Response, and Cancer in Aging Telomerase-Deficient Mice." *Cell* 96 (5): 701–12.

Ruley, H. E. 1983. "Adenovirus Early Region 1A Enables Viral and Cellular Transforming Genes to Transform Primary Cells in Culture." *Nature* 304 (5927): 602–6.

Salvioli, S., M. Capri, L. Bucci, C. Lanni, M. Racchi, D. Uberti, M. Memo, D. Mari, S. Govoni, and C. Franceschi. 2009. "Why Do Centenarians Escape or Postpone Cancer? The Role of IGF-1, Inflammation and p53." *Cancer Immunology, Immunotherapy* 58 (12): 1909–17.

Sanoff, H. K., A. M. Deal, J. Krishnamurthy, C. Torrice, P. Dillon, J. Sorrentino, J. G. Ibrahim, et al. 2014. "Effect of Cytotoxic Chemotherapy on Markers of Molecular Age in Patients with Breast Cancer." *Journal of the National Cancer Institute* 106 (4): dju057.

Scadden, D. T. 2014. "Nice Neighborhood: Emerging Concepts of the Stem Cell Niche." *Cell* 157 (1): 41–50.

Sekine, Y., A. Hata, E. Koh, and K. Hiroshima. 2014. "Lung Carcinogenesis from Chronic Obstructive Pulmonary Disease: Characteristics of Lung Cancer from COPD and Contribution of Signal Transducers and Lung Stem Cells in the Inflammatory Microenvironment." *General Thoracic and Cardiovascular Surgery* 62 (7): 415–21.

Seluanov, A., Z. Chen, C. Hine, T. H. C. Sasahara, A. A. C. M. Ribeiro, K. C. Catania, D. C. Presgraves, and V. Gorbunova. 2007. "Telomerase Activity Coevolves with Body Mass, Not Lifespan." *Aging Cell* 6 (1): 45–52.

Sequist, L. V., B. A. Waltman, D. Dias-Santagata, S. Digumarthy, A. B. Turke, P. Fidias, K. Bergethon, et al. 2011. "Genotypic and Histological Evolution of Lung Cancers Acquiring Resistance to EGFR Inhibitors." *Science Translational Medicine* 3 (75): 75ra26.

Serrano, M., and M. A. Blasco. 2007. "Cancer and Ageing: Convergent and Divergent Mechanisms." *Nature Reviews Molecular Cell Biology* 8 (9): 715–22.

Serrano, M., A. W. Lin, M. E. McCurrach, D. Beach, and S. W. Lowe. 1997. "Oncogenic *ras* Provokes Premature Cell Senescence Associated with Accumulation of p53 and p16INK4a." *Cell* 88 (5): 593–602.

Seymour, R. S., V. Bosiocic, and E. P. Snelling. 2016. "Fossil Skulls Reveal That Blood Flow Rate to the Brain Increased Faster Than Brain Volume during Human Evolution." *Royal Society Open Science* 3: 160305.

Shao, L., Y. Luo, and D. Zhou. 2014. "Hematopoietic Stem Cell Injury Induced by Ionizing Radiation." *Antioxidants and Redox Signaling* 20 (9): 1447–62.

Sharma, G., N. A. Hanania, and Y. M. Shim. 2009. "The Aging Immune System and Its Relationship to the Development of Chronic Obstructive Pulmonary Disease." *Proceedings of the American Thoracic Society* 6 (7): 573–80.

Sharpless, N. E., and R. A. DePinho. 2007. "How Stem Cells Age and Why This Makes Us Grow Old." *Nature Reviews Molecular Cell Biology* 8 (9): 703–13.

Shelford, V. E. 1931. "Some Concepts in Bioecology." *Ecology* 12 (3): 455–67.

Siegel, R., D. Naishadham, and A. Jemal. 2012. Cancer statistics, 2012. *CA: a cancer journal for clinicians* 62(1): 10–29.

Signer, R. A. J., J. A. Magee, A. Salic, and S. J. Morrison. 2014. "Haematopoietic Stem Cells Require a Highly Regulated Protein Synthesis Rate." *Nature* 509 (7498): 49–54.

Simons, M. J. 2015. "Questioning Causal Involvement of Telomeres in Aging." *Ageing Research Reviews* 24 (Part B): 191–96.

Simpson, G. G. 1944. *Tempo and Mode in Evolution.* New York: Columbia University Press.

Smith, G. S., R. L. Walford, and M. R. Mickey. 1973. "Lifespan and Incidence of Cancer and Other Diseases in Selected Long-Lived Inbred Mice and Their F1 Hybrids." *Journal of the National Cancer Institute* 50 (5): 1195–213.

Smith, K. R., H. A. Hanson, G. P. Mineau, and S. S. Buys. 2012. "Effects of BRCA1 and BRCA2 Mutations on Female Fertility." *Proceedings of the Royal Society B: Biological Sciences* 279 (1732): 1389–95.

Snippert, H. J., L. G. van der Flier, T. Sato, J. H. van Es, M. van den Born, C. Kroon-Veenboer, N. Barker, et al. 2010. "Intestinal Crypt Homeostasis Results from Neutral Competition between Symmetrically Dividing Lgr5 Stem Cells." *Cell* 143 (1): 134–44.

Soto, A. M., and C. Sonnenschein. 2004. "The Somatic Mutation Theory of Cancer: Growing Problems with the Paradigm?" *BioEssays* 26 (10): 1097–107.

Stein, C. J., and G. A. Colditz. 2004. "Modifiable Risk Factors for Cancer." *British Journal of Cancer* 90 (2): 299–303.

Stewart, T. A., and B. Mintz. 1981. "Successive Generations of Mice Produced from an Established Culture Line of Euploid Teratocarcinoma Cells." *Proceedings of the National Academy of Sciences of the United States of America* 78 (10): 6314–18.

Strassmann, B. I. 1999. "Menstrual Cycling and Breast Cancer: An Evolutionary Perspective." *Journal of Women's Health* 8 (2): 193–202.

Sulak, M., L. Fong, K. Mika, S. Chigurupati, L. Yon, N. P. Mongan, R. D. Emes, and V. J. Lynch. 2016. "TP53 Copy Number Expansion Is Associated with the Evolution of Increased Body Size and an Enhanced DNA Damage Response in Elephants." *eLife* 5: e11994.

Sutter, N. B., C. D. Bustamante, K. Chase, M. M. Gray, K. Zhao, L. Zhu, B. Padhukasahasram, et al. 2007. "A Single IGF1 Allele Is a Major Determinant of Small Size in Dogs." *Science* 316 (5821): 112–15.

Takahashi, H., H. Ogata, R. Nishigaki, D. H. Broide, and M. Karin. 2010. "Tobacco Smoke Promotes Lung Tumorigenesis by Triggering IKKbeta- and JNK1-Dependent Inflammation." *Cancer Cell* 17 (1): 89–97.

Takamura, A., M. Komatsu, T. Hara, A. Sakamoto, C. Kishi, S. Waguri, Y. Eishi, O. Hino, K. Tanaka, and N. Mizushima. 2011. "Autophagy-Deficient Mice Develop Multiple Liver Tumors." *Genes and Development* 25 (8): 795–800.

Takebe, N., L. Miele, P. J. Harris, W. Jeong, H. Bando, M. Kahn, S. X. Yang, and S. P. Ivy. 2015. "Targeting Notch, Hedgehog, and Wnt Pathways in Cancer Stem Cells: Clinical Update." *Nature Reviews Clinical Oncology* 12 (8): 445–64.

Takiguchi, Y., I. Sekine, S. Iwasawa, R. Kurimoto, and K. Tatsumi. 2014. "Chronic Obstructive Pulmonary Disease as a Risk Factor for Lung Cancer." *World Journal of Clinical Oncology* 5 (4): 660–66.

Taub, R. 2004. "Liver Regeneration: From Myth to Mechanism." *Nature Reviews Molecular Cell Biology* 5 (10): 836–47.

Thomas, F., R. M. Nesse, R. Gatenby, C. Gidoin, F. Renaud, B. Roche, and B. Ujvari. 2016. "Evolutionary Ecology of Organs: A Missing Link in Cancer Development?" *Trends in Cancer* 2 (8): 409–15.

Thomas, F., B. Ujvari, C. Gidouin, A. Tasiemski, P. W. Ewald, and B. Roche. 2017. "Toward an Ultimate Explanation of Intratumor Heterogeneity." In *Ecology and Evolution of Cancer*, edited by B. Ujvari, B. Roche, and F. Thomas, 219–22. London: Academic Press.

Thompson, C. B., D. E. Bauer, J. J. Lum, G. Hatzivassiliou, W. X. Zong, F. Zhao, D. Ditsworth, M. Buzzai, and T. Lindsten. 2005. "How Do Cancer Cells Acquire the Fuel Needed to Support Cell Growth?" *Cold Spring Harbor Symposia on Quantitative Biology* 70: 357–62.

Tian, X., J. Azpurua, C. Hine, A. Vaidya, M. Myakishev-Rempel, J. Ablaeva, Z. Mao, E. Nevo, V. Gorbunova, and A. Seluanov. 2013. "High-Molecular-Mass Hyaluronan Mediates the Cancer Resistance of the Naked Mole Rat." *Nature* 499 (7458): 346–49.

Tishkoff, S. A., F. A. Reed, A. Ranciaro, B. F. Voight, C. C. Babbitt, J. S. Silverman, K. Powell, et al. 2007. "Convergent Adaptation of Human Lactase Persistence in Africa and Europe." *Nature Genetics* 39 (1): 31–40.

Todoric, J., L. Antonucci, and M. Karin. 2016. "Targeting Inflammation in Cancer Prevention and Therapy." *Cancer Prevention Research* 9 (12): 895–905.

Toft, N. J., D. J. Winton, J. Kelly, L. A. Howard, M. Dekker, H. te Riele, M. J. Arends, A. H. Wyllie, G. P. Margison, and A. R. Clarke. 1999. "Msh2 Status Modulates Both Apoptosis and Mutation Frequency in the Murine Small Intestine." *Proceedings of the National Academy of Sciences of the United States of America* 96 (7): 3911–15.

Tomasetti, C., L. Li, and B. Vogelstein. 2017. "Stem Cell Divisions, Somatic Mutations, Cancer Etiology, and Cancer Prevention." *Science* 355 (6331): 1330–34.

Tomasetti, C., L. Marchionni, M. A. Nowak, G. Parmigiani, and B. Vogelstein. 2015. "Only Three Driver Gene Mutations Are Required for the Development of Lung and Colorectal Cancers." *Proceedings of the National Academy of Sciences of the United States of America* 112 (1): 118–23.

Tomasetti, C., and B. Vogelstein. 2015. "Cancer Etiology. Variation in Cancer Risk among Tissues Can Be Explained by the Number of Stem Cell Divisions." *Science* 347 (6217): 78–81.

Tuljapurkar, S. D., C. O. Puleston, and M. D. Gurven. 2007. "Why Men Matter: Mating Patterns Drive Evolution of Human Lifespan." *PLoS One* 2 (8): e785.

Tzu, S. 2013. *The Art of War.* Translated by Lionel Giles. Bronx, NY: Ishi Press International.

Ujvari, B., R. A. Gatenby, and F. Thomas. 2016. "The Evolutionary Ecology of Transmissible Cancers." *Infection, Genetics and Evolution* 39: 293–303.

UK Office for National Statistics. n.d. Accessed September 3, 2017. https://www.ons.gov.uk/peoplepopulationandcommunity/birthsdeaths andmarriages/lifeexpectancies/datasets/nationallifetablesgreatbritain referencetables.

Vander Heiden, M. G., L. C. Cantley, and C. B. Thompson. 2009. "Understanding the Warburg effect: the metabolic requirements of cell proliferation." *Science* 324(5930): 1029–33.

Van der Put, E., D. Frasca, A. M. King, B. B. Blomberg, and R. L. Wiley. 2004. "Decreased E47 in Senescent B Cell Precursors Is Stage Specific and Regulated Posttranslationally by Protein Turnover." *Journal of Immunology* 173 (2): 818–27.

Vas, V., K. Senger, K. Dörr, A. Niebel, and H. Geiger. 2012. "Aging of the Microenvironment Influences Clonality in Hematopoiesis." *PLoS One* 7 (8): e42080.

Vas, V., C. Wandhoff, K. Dörr, A. Niebel, and H. Geiger. 2012. "Contribution of an Aged Microenvironment to Aging-Associated Myeloproliferative Disease." *PLoS One* 7 (2): e31523.

Venkatesan, R. N., P. M. Treuting, E. D. Fuller, R. E. Goldsby, T. H. Norwood, T. A. Gooley, W. C. Ladiges, B. D. Preston, and L. A. Loeb. 2007. "Mutation at the Polymerase Active Site of Mouse DNA Polymerase Delta Increases Genomic Instability and Accelerates Tumorigenesis." *Molecular and Cellular Biology* 27 (21): 7669–82.

Vermeulen, L., E. Morrissey, M. van der Heijden, A. M. Nicholson, A. Sottoriva, S. Buczacki, R. Kemp, S. Tavaré, and D. J. Winton. 2013. "Defining Stem Cell Dynamics in Models of Intestinal Tumor Initiation." *Science* 342 (6161): 995–98.

Vijg, J., R. A. Busuttil, R. Bahar, and M. E. Dollé. 2005. "Aging and Genome Maintenance." *Annals of the New York Academy of Sciences* 1055: 35–47.

Vittecoq, M., H. Ducasse, A. Arnal, A. P. Møller, B. Ujvari, C. B. Jacqueline, T. Tissot, et al. 2015. "Animal Behaviour and Cancer." *Animal Behaviour* 101: 19–26.

Vousden, K. H., and D. P. Lane. 2007. "p53 in Health and Disease." *Nature Reviews Molecular Cell Biology* 8 (4): 275–83.

Waddington, C. H. 1957. *The Strategy of the Genes: A Discussion of Some Aspects of Theoretical Biology.* London: Allen and Unwin.

Wahl, G. M., and B. T. Spike. 2017. "Cell State Plasticity, Stem Cells, EMT, and the Generation of Intra-Tumoral Heterogeneity." *NPJ Breast Cancer* 3 (1): 14.

Waldhauer, I., and A. Steinle. 2008. "NK Cells and Cancer Immunosurveillance." *Oncogene* 27 (45): 5932–43.

Wang, J. C., M. Doedens, and J. E. Dick. 1997. "Primitive Human Hematopoietic Cells Are Enriched in Cord Blood Compared with Adult Bone Marrow or Mobilized Peripheral Blood as Measured by the Quantitative In Vivo SCID-Repopulating Cell Assay." *Blood* 89 (11): 3919–24.

Ward, P. D., and J. L. Kirschvink. 2015. *A New History of Life: The Radical New Discoveries about the Origins and Evolution of Life on Earth.* New York: Bloomsbury Press.

Weinreich, D. M., N. F. Delaney, M. A. Depristo, and D. L. Hartl. 2006. "Darwinian Evolution Can Follow Only Very Few Mutational Paths to Fitter Proteins." *Science* 312 (5770): 111–14.

West, H., G. R. Oxnard, and R. C. Doebele. 2013. "Acquired Resistance to Targeted Therapies in Advanced Non-Small Cell Lung Cancer: New Strategies and New Agents." American Society of Clinical Oncology Educational Book. Accessed September 3, 2017. doi: 10.1200/EdBook _AM.2013.33.e272.

Westrich, J. A., C. J. Warren, and D. Pyeon. 2017. "Evasion of Host Immune Defenses by Human Papillomavirus." *Virus Research* 231: 21–33.

White, R. R., and J. Vijg. 2016. "Do DNA Double-Strand Breaks Drive Aging?" *Molecular Cell* 63 (5): 729–38.

Wierzbicki, A. S., and A. Viljoen. 2010. "Hyperlipidaemia in Paediatric Patients: The Role of Lipid-Lowering Therapy in Clinical Practice." *Drug Safety* 33 (2): 115–25.

Wilde, O. 1891. *The Picture of Dorian Gray.* London: Ward, Lock and Co.

Williams, G. C. 1957. "Pleiotropy, Natural Selection, and the Evolution of Senescence." *Evolution* 11 (4): 398–411.

——. 1966. *Adaptation and Natural Selection: A Critique of Some Current Evolutionary Thought.* Princeton, NJ: Princeton University Press.

Wong, T. N., G. Ramsingh, A. L. Young, C. A. Miller, W. Touma, J. S. Welch, T. L. Lamprecht, et al. 2015. "Role of TP53 Mutations in the Origin and Evolution of Therapy-Related Acute Myeloid Leukaemia." *Nature* 518 (7540): 552–55.

Wrangham, R., and R. Carmody. 2010. "Human Adaptation to the Control of Fire." *Evolutionary Anthropology* 19 (5): 187–99.

Wright, D. E., A. J. Wagers, A. P. Gulati, F. L. Johnson, and I. L. Weissman. 2001. "Physiological Migration of Hematopoietic Stem and Progenitor Cells." *Science* 294 (5548): 1933–36.

Wright, S. 1931. "Evolution in Mendelian Populations." *Genetics* 16 (2): 97–159.

———. 1932. "The Roles of Mutation, Inbreeding, Crossbreeding and Selection in Evolution." In *Proceedings of the Sixth International Congress on Genetics*. Accessed March 2, 2012. http://www.blackwellpublishing.com /ridley/classictexts/wright.pdf, 355–66.

Wu, C. Y., M. S. Wu, K. N. Kuo, C. B. Wang, Y. J. Chen, and J. T. Lin. 2010. "Effective Reduction of Gastric Cancer Risk with Regular Use of Nonsteroidal Anti-Inflammatory Drugs in Helicobacter Pylori-Infected Patients." *Journal of Clinical Oncology* 28 (18): 2952–57.

Wu, S., S. Powers, W. Zhu, and Y. A. Hannun. 2016. "Substantial Contribution of Extrinsic Risk Factors to Cancer Development." *Nature* 529 (7584): 43–47.

Xie, M., C. Lu, J. Wang, M. D. McLellan, K. J. Johnson, M. C. Wendl, J. F. McMichael, et al. 2014. "Age-Related Mutations Associated with Clonal Hematopoietic Expansion and Malignancies." *Nature Medicine* 20 (12): 1472–78.

Xue, W., L. Zender, C. Miething, R. A. Dickins, E. Hernando, V. Krizhanovsky, C. Cordon-Cardo, and S. W. Lowe. 2007. "Senescence and Tumour Clearance Is Triggered by p53 Restoration in Murine Liver Carcinomas." *Nature* 445 (7128): 656–60.

Yadav, V. K., J. DeGregori, and S. De. 2016. "The Landscape of Somatic Mutations in Protein Coding Genes in Apparently Benign Human Tissues Carries Signatures of Relaxed Purifying Selection." *Nucleic Acids Research* 44 (5): 2075–84.

Yates, L. R., and P. J. Campbell. 2012. "Evolution of the Cancer Genome." *Nature Reviews Genetics* 13 (11): 795–806.

Zhang, C., Y. Guan, Y. Sun, D. Ai, and Q. Guo. 2016. "Tumor Heterogeneity and Circulating Tumor Cells." *Cancer Letters* 374 (2): 216–23.

Zhao, Z. M., B. Zhao, Y. Bai, A. Iamarino, S. G. Gaffney, J. Schlessinger, R. P. Lifton, D. L. Rimm, and J. P. Townsend. 2016. "Early and Multiple

Origins of Metastatic Lineages within Primary Tumors." *Proceedings of the National Academy of Sciences of the United States of America* 113 (8): 2140–45.

Ziv, O., B. Glaser, and Y. Dor. 2013. "The Plastic Pancreas." *Developmental Cell* 26 (1): 3–7.

Acknowledgments

Ideas do not form in a vacuum, and I have greatly benefited from teachers, mentors, colleagues, students, postdoctoral fellows, and friends over the years. Ida Medlen, my biology teacher at Bellaire High School, solidified my interest in biology and disease—she made science exciting. I enjoyed a life-changing experience in the town of Tomás Gomensoro in Uruguay for one year as an exchange student, hosted by the Ramos family. This experience stimulated my adventurous side and sadly also resulted in my first encounter with the tragic consequences of cancer, as my host father Atilio Ramos succumbed to leukemia a few years after I returned to the United States. At the University of Texas, I got my first full experience with research under the guidance of Henry (Hank) Bose—I was hooked on research from that point forward. My graduate years at the Massachusetts Institute of Technology, where I worked under the mentorship of H. Earl Ruley, further forged my critical thinking skills and research prowess. At Duke University, I benefited from the excellent guidance of Joseph Nevins. I also want to thank Niles Eldredge, a scientific hero of mine, for enlightening e-mail conversations and, with Stephen Jay Gould, for developing the theory of punctuated evolution that has greatly influenced my ideas on cancer.

I want to express my gratitude to Janice Audet, my editor at Harvard University Press, for spurring me to write this book, and for her careful editing and critical comments, and to her editorial assistant Emeralde Jensen-Roberts for ushering the manuscript through the various stages. Jeanne

Ferris at Westchester Publishing Services masterfully copy-edited the book, which substantially improved readability and consistency. I also received excellent suggestions for changes from Paul Ewald and an anonymous reviewer.

My own laboratory has played a large role in the development of the theory of adaptive oncogenesis. Early work by Feng Li and Ganna Bilousova provided the first experimental support for the theory, followed by critical studies and theoretical development with Andriy Marusyk. Curtis Henry supplied a mechanism—inflammation—to account for the altered fitness landscapes in the aged bone marrow, a critical insight that suggests tissue landscapes can be modulated. Courtney Fleenor and Kelly Higa provided novel mechanistic insights into oncogenic adaptation within bone marrow stem cell pools permanently damaged by radiation exposure. Andrii Rozhok's computational modeling gave mathematical substantiation for adaptive oncogenesis, and his ideas further strengthened its theoretical support. I am grateful to Hannah Scarborough for critically evaluating many aspects of the theory and for providing insightful and very useful comments on this book. I have also greatly benefited from discussions with many of my colleagues over the years, including (but not limited to) Robert Sclafani, Heide Ford, Mark Johnston, Elan Eisenmesser, Charles Dinarello, Paul Bunn, Lynn Heasley, Ruth Hershberg, Craig Jordan, Andrew Thorburn, Robert Gatenby, Angelika Amon, Hartmut Geiger, Geoffrey Wahl, and Subhajyoti De.

Scientific research is not cheap, and the development of the theory of adaptive oncogenesis necessitated complex and expensive experiments. Funding from a number of agencies has been key, including from the National Cancer Institute, Cancer League of Colorado, Leukemia and Lymphoma Society, and V Foundation. In recognition of the critical role of private foundations in supporting cancer research, I am donating all of the proceeds from this book to the Leukemia and Lymphoma Society and the V Foundation.

Much of this book was written while I was on sabbatical at the University of Vermont in Burlington. Many thanks to my colleagues there for providing me with an intellectually stimulating environment. Mercedes Rincon hosted me for the five months in Vermont. She not only provided her love and support, but also applied her impressive and inquisitive mind to this project.

My sons, Luke and Michael DeGregori, have always been an inspiration to me, and I have bounced many ideas off them. Michael is also the creator of the wonderful artwork in this book, adding needed character and at times levity to a very serious topic. Finally, I want to thank my parents, Thomas and Gayle DeGregori, for their lifelong support. They have always believed in me, stressed the importance of a critical mind, and given me the freedom to explore the world around me.

Index